LONDON MATHEMATICAL SOCIETY LECTURE NOTE SERIES

Managing Editor: Professor J.W.S. Cassels, Department of Pure Mathematics and Mathematical Statistics, 16 Mill Lane, Cambridge CB2 1SB, England

4 Algebraic topology, J.F.ADAMS
5 Commutative algebra, J.T.KNIGHT
8 Integration and harmonic analysis on compact groups, R.E.EDWARDS
11 New developments in topology, G.SEGAL (ed)
12 Symposium on complex analysis, J.CLUNIE & W.K.HAYMAN (eds)
13 Combinatorics, T.P.McDONOUGH & V.C.MAVRON (eds)
15 An introduction to topological groups, P.J.HIGGINS
16 Topics in finite groups, T.M.GAGEN
17 Differential germs and catastrophes, Th.BROCKER & L.LANDER
18 A geometric approach to homology theory, S.BUONCRISTIANO, C.P.ROURKE & B.J.SANDERSON
20 Sheaf theory, B.R.TENNISON
21 Automatic continuity of linear operators, A.M.SINCLAIR
23 Parallelisms of complete designs, P.J.CAMERON
24 The topology of Stiefel manifolds, I.M.JAMES
25 Lie groups and compact groups, J.F.PRICE
26 Transformation groups, C.KOSNIOWSKI (ed)
27 Skew field constructions, P.M.COHN
29 Pontryagin duality and the structure of LCA groups, S.A.MORRIS
30 Interaction models, N.L.BIGGS
31 Continuous crossed products and type III von Neumann algebras,A.VAN DAELE
32 Uniform algebras and Jensen measures, T.W.GAMELIN
34 Representation theory of Lie groups, M.F. ATIYAH <u>et al.</u>
35 Trace ideals and their applications, B.SIMON
36 Homological group theory, C.T.C.WALL (ed)
37 Partially ordered rings and semi-algebraic geometry, G.W.BRUMFIEL
38 Surveys in combinatorics, B.BOLLOBAS (ed)
39 Affine sets and affine groups, D.G.NORTHCOTT
40 Introduction to Hp spaces, P.J.KOOSIS
41 Theory and applications of Hopf bifurcation, B.D.HASSARD, N.D.KAZARINOFF & Y-H.WAN
42 Topics in the theory of group presentations, D.L.JOHNSON
43 Graphs, codes and designs, P.J.CAMERON & J.H.VAN LINT
44 Z/2-homotopy theory, M.C.CRABB
45 Recursion theory: its generalisations and applications, F.R.DRAKE & S.S.WAINER (eds)
46 p-adic analysis: a short course on recent work, N.KOBLITZ
47 Coding the Universe, A.BELLER, R.JENSEN & P.WELCH
48 Low-dimensional topology, R.BROWN & T.L.THICKSTUN (eds)
49 Finite geometries and designs,P.CAMERON, J.W.P.HIRSCHFELD & D.R.HUGHES (eds)
50 Commutator calculus and groups of homotopy classes, H.J.BAUES
51 Synthetic differential geometry, A.KOCK
52 Combinatorics, H.N.V.TEMPERLEY (ed)
53 Singularity theory, V.I.ARNOLD
54 Markov process and related problems of analysis, E.B.DYNKIN
55 Ordered permutation groups, A.M.W.GLASS
56 Journêes arithmêtiques, J.V.ARMITAGE (ed)
57 Techniques of geometric topology, R.A.FENN
58 Singularities of smooth functions and ma
59 Applicable differential geometry, M.CRAM
60 Integrable systems, S.P.NOVIKOV <u>et al</u>
61 The core model, A.DODD

62 Economics for mathematicians, J.W.S.CASSELS
63 Continuous semigroups in Banach algebras, A.M.SINCLAIR
64 Basic concepts of enriched category theory, G.M.KELLY
65 Several complex variables and complex manifolds I, M.J.FIELD
66 Several complex variables and complex manifolds II, M.J.FIELD
67 Classification problems in ergodic theory, W.PARRY & S.TUNCEL
68 Complex algebraic surfaces, A.BEAUVILLE
69 Representation theory, I.M.GELFAND et al.
70 Stochastic differential equations on manifolds, K.D.ELWORTHY
71 Groups - St Andrews 1981, C.M.CAMPBELL & E.F.ROBERTSON (eds)
72 Commutative algebra: Durham 1981, R.Y.SHARP (ed)
73 Riemann surfaces: a view towards several complex variables,A.T.HUCKLEBERRY
74 Symmetric designs: an algebraic approach, E.S.LANDER
75 New geometric splittings of classical knots, L.SIEBENMANN & F.BONAHON
76 Linear differential operators, H.O.CORDES
77 Isolated singular points on complete intersections, E.J.N.LOOIJENGA
78 A primer on Riemann surfaces, A.F.BEARDON
79 Probability, statistics and analysis, J.F.C.KINGMAN & G.E.H.REUTER (eds)
80 Introduction to the representation theory of compact and locally compact groups, A.ROBERT
81 Skew fields, P.K.DRAXL
82 Surveys in combinatorics, E.K.LLOYD (ed)
83 Homogeneous structures on Riemannian manifolds, F.TRICERRI & L.VANHECKE
84 Finite group algebras and their modules, P.LANDROCK
85 Solitons, P.G.DRAZIN
86 Topological topics, I.M.JAMES (ed)
87 Surveys in set theory, A.R.D.MATHIAS (ed)
88 FPF ring theory, C.FAITH & S.PAGE
89 An F-space sampler, N.J.KALTON, N.T.PECK & J.W.ROBERTS
90 Polytopes and symmetry, S.A.ROBERTSON
91 Classgroups of group rings, M.J.TAYLOR
92 Representation of rings over skew fields, A.H.SCHOFIELD
93 Aspects of topology, I.M.JAMES & E.H.KRONHEIMER (eds)
94 Representations of general linear groups, G.D.JAMES
95 Low-dimensional topology 1982, R.A.FENN (ed)
96 Diophantine equations over function fields, R.C.MASON
97 Varieties of constructive mathematics, D.S.BRIDGES & F.RICHMAN
98 Localization in Noetherian rings, A.V.JATEGAONKAR
99 Methods of differential geometry in algebraic topology, M.KAROUBI & C.LERUSTE
100 Stopping time techniques for analysts and probabilists, L.EGGHE
101 Groups and geometry, ROGER C.LYNDON
102 Topology of the automorphism group of a free group, S.M.GERSTEN
103 Surveys in combinatorics 1985, I.ANDERSEN (ed)
104 Elliptical structures on 3-manifolds, C.B.THOMAS
105 A local spectral theory for closed operators, I.ERDELYI & WANG SHENGWANG
106 Syzygies, E.G.EVANS & P.GRIFFITH
107 Compactification of Siegel moduli schemes, C-L.CHAI
108 Some topics in graph theory, H.P.YAP
109 Diophantine analysis, J.LOXTON & A.VAN DER POORTEN (eds)

London Mathematical Society Lecture Note Series. 108

Some Topics in Graph Theory

H.P. YAP
National University of Singapore

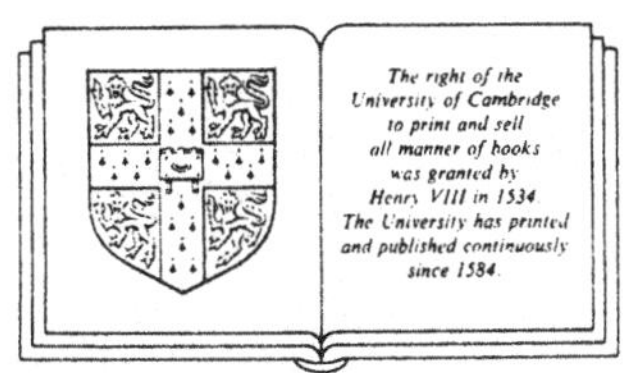

CAMBRIDGE UNIVERSITY PRESS
Cambridge
New York New Rochelle
Melbourne Sydney

CAMBRIDGE UNIVERSITY PRESS
Cambridge, New York, Melbourne, Madrid, Cape Town, Singapore, São Paulo

Cambridge University Press
The Edinburgh Building, Cambridge CB2 8RU, UK

Published in the United States of America by Cambridge University Press, New York

www.cambridge.org
Information on this title: www.cambridge.org/9780521339445

First published 1986
Reprinted 1987
Re-issued in this digitally printed version 2008

A catalogue record for this publication is available from the British Library

Library of Congress Cataloguing in Publication data

Yap, H.P. (Hian Poh), 1938–
Some topics in graph theory. (London Mathematical Society
lecture note series; 108)
Includes bibliographies and index.
1. Graph theory.
I. Title. II. Series.
QA166.Y37 1986 511'.5 86–9548

ISBN 978-0-521-33944-5 paperback

Introduction

A complete, optional course on Graph Theory was first offered to Fourth Year Honours students of the Department of Mathematics, National University of Singapore in the academic year 1982/83. To those students taking this course, it was their first introduction to Graph Theory and so the standard of the course could not be set too high. However, since it was a fourth year Honours Course, the standard could not be too low. For this reason, I decided to use some existing textbooks for the basic results in the first term and to concentrate on only a few special topics in the second term in order to expose the students to some very recent results. This book eventually grew out from the lectures I gave to the students during the academic years 1982/83 and 1983/84.

More than seventy per cent of the materials in this book are taken directly from recent research papers. Each chapter (except chapter 1) gives an up-to-date account of a particular topic in Graph Theory which is very active in current research. In addition, detail of proofs of all the theorems are given and numerous exercises and open problems are included. Thus this book is not only suitable for use as a supplement to a course text at advanced undergraduate or postgraduate level, but will also, I hope, be of some help to researchers in Graph Theory. In fact, Mr. Chen Jing-Hui had written to inform me that by using my lecture notes in his fourth year Graph Theory course in Xiamen University, his students were able to do some research straightway.

A preliminary draft of the manuscript consisting of most of the sections of Chapters 2, 3 and 5 was first submitted to Professor E. B. Davies for consideration to be published as a volume in the London Mathematical Society Student Texts Series in June 1983. This draft was refereed and subsequently transferred to Professor J. W. S. Cassels for consideration to be published as a volume in the London Mathematical Society Lecture Note Series. I am very grateful to the referees for their valuable comments and constructive criticisms, especially for providing a shorter proof of Turner's results on vertex-transitive graphs of prime order. I am also very grateful to Professors J. W. S. Cassels and E. B. Davies for their encouragement. Thanks are also due to Professor J. W. S. Cassels for assisting me in my application to the National University of Singapore for a 5-month (November 19, 1984 to April 18, 1985) sabbatical leave and to the National University of Singapore for approving my application so that I can concentrate working on this project.

While I was writing this book, I had opportunities to give several survey talks on some topics covered in the book to various institutions in England and China. In November 1982, I visited Oxford University, University of Birmingham and the Open University for two weeks under the sponsorship of the British Council. In May 1984, while on a sightseeing tour of China, I visited East China Normal University (Shanghai), University of Science and Technology (Hefei) and Academia Sinica (Beijing). I also visited University of Cambridge, University of Reading, Oxford University and the Royal Holloway College in January/ February, 1985 while I was on sabbatical leave. It is now a great pleasure for me to acknowledge the helpful comments and suggestions from many friends including N. L. Biggs, B. Bollobás, P. J. Cameron, Dong Chun-Fei, A. D. Gardiner, A. J. W. Hilton, E. C. Milner, Wang Jian-Fang, D. J. A. Welsh and R. J. Wilson. I am also indebted to Professor R. M. Weiss for providing a sketch of his nice proof of Tutte's theorem on s-transitive cubic graphs and to my colleague K. M. Koh for helping me in the proofreading. Miss D. Shanthi's excellent word-processing should also be recorded here.

Finally, a few words on the reference system and the exercises of this book. When Theorem i.j is referred to, unless otherwise specified, it is meant that we are referring to Theorem i.j of the same chapter. When an exercise is marked with a minus or a plus sign, it means that the exercise is easy or hard/time-consuming respectively; and if it is marked with a star, it means that it is an open problem or a conjecture.

Contents

1. BASIC TERMINOLOGY

1. Basic graph-theoretic terms

In this section we shall define some basic graph-theoretic terms that will be used in this book. Other graph-theoretic terms which are not included in this section will be defined when they are needed.

Unless stated otherwise, all graphs are finite, undirected, simple and loopless. A directed graph is called a digraph and a directed edge is called an arc. A multigraph permits more than one edge joining two of its vertices. The number of edges joining two vertices u and v is called the multiplicity of uv and is denoted by $\mu(u,v)$.

The cardinality of a set S is denoted by $|S|$. Let $G = (V,E)$ be a graph where $V = V(G)$ is its vertex set and $E = E(G)$ is its edge set. The order (resp. size) of G is $|V|$ (resp. $|E|$) and is denoted by $|G|$ (resp. $e(G)$). Two vertices u and v of G are said to be adjacent if $uv \in E$. If $e = uv \in E$, then we say that u and v are the end-vertices of e and that the edge e is incident with u and v. Two edges e and f of G are said to be adjacent if they have one common end-vertex. If $uv \in E$, then we say that v is a neighbour of u. The set of all neighbours of u is called the neighbourhood of u and is denoted by $N_G(u)$ or simply by $N(u)$ if there is no danger of confusion. The valency (or degree) of a vertex u is $|N(u)|$ and is denoted by $d(u)$. The maximum (resp. minimum) of the valencies of the vertices of G is called the maximum (resp. mininum) valency of G and is denoted by $\Delta(G)$ (resp. $\delta(G)$).

A graph H is said to be a subgraph of a graph G if $V(H) \subseteq V(G)$ and $E(H) \subseteq E(G)$. A subgraph H of G such that whenever $u, v \in V(H)$ are adjacent in G then they are also adjacent in H is called an induced subgraph of G. An induced subgraph of G having vertex set (or a subgraph of G induced by) $\{v_1, \ldots, v_k\}$ is denoted by $\langle v_1, \ldots, v_k \rangle$. The subgraph induced by $V(G) - \{v_1, \ldots, v_k\}$ is denoted by $G - \{v_1, \ldots, v_k\}$ or by $G - v_1 - \ldots - v_k$.

A vertex of valency 0 is called an isolated vertex. If all the vertices of G have the same valency, d say, then we say that G is regular of degree d and we write deg G = d. A regular graph of degree 3 is called a cubic graph. If G is a regular graph of order n such that deg G = 0 (resp. n - 1), then G is called a null graph (resp. complete graph) and is denoted by 0_n (resp. K_n). If the vertex set of G can be partitioned into two sets V_1 and V_2 such that every edge of G joins one vertex in V_1 to one vertex in V_2, then G is called a bipartite graph. A bipartite graph having bipartition V_1 and V_2 is said to be complete if each vertex in V_1 is adjacent to every vertex in V_2. The complete bipartite graph having bipartition V_1 and V_2 such that $|V_1| = r$ and $|V_2| = s$ is denoted by $K_{r,s}$. A complete bipartite graph $K_{1,r}$ is called a star and is denoted by S_{r+1}. The Petersen graph G(5,2) is a cubic graph having vertex set $V = \{u_0, \ldots, u_4, v_0, \ldots, v_4\}$ and edge set $E = \{(u_i, u_{i+1}), (u_i, v_i), (v_i, v_{i+2}) \mid i = 0, \ldots, 4\}$ where all the subscripts are taken modulo 5. The generalized Petersen graph G(n,k) ($n > 5$, $0 < k < n$) is the cubic graph having vertex set $\{u_0, \ldots, u_{n-1}, v_0, \ldots, v_{n-1}\}$ and edge set $\{(u_i, u_{i+1}), (u_i, v_i), (v_i, v_{i+k}) \mid i = 0, \ldots, n - 1\}$ where all the subscripts are taken modulo n. The graphs whose vertices and edges are the vertices and edges of the five regular solids are called the platonic graphs.

An independent set of edges, or matching, in G is a set of edges no two of which are adjacent. A matching in G that includes every vertex of G is called a 1-factor in G.

A sequence of distinct edges of the form $v_0 v_1, v_1 v_2, \ldots, v_{r-1} v_r$ is called a path of length r from v_0 to v_r. If the vertices $v_0, v_1, \ldots, v_r$ are all distinct, then the path is called a chain (or open chain), whereas if the vertices are all distinct except that $v_r = v_0$, then the path is a cycle (or circuit). The length of a shortest open chain from a vertex u to a vertex $v \neq u$ is called the distance between u and v and is denoted by $\partial(u,v)$. The maximum distance between two vertices of G is called the diameter of G and is denoted by d(G). The length of a shortest cycle in G is called the girth of G and is denoted by $\gamma(G)$. The length of a longest cycle in G is called the circumference of G. A cycle of length n is denoted by C_n and a shortest path (open chain) of length n is denoted by P_n. If G has an open chain P that includes every vertex of G, then P is called a Hamilton path (or H-path) of G. A cycle that includes all the vertices of G is called a Hamilton cycle of G. If G has a Hamilton cycle, then G is said to be Hamiltonian.

Two graphs G and H are said to be disjoint if they have no vertex in common. Suppose G and H are two disjoint graphs. Then the (disjoint) union $G \cup H$ of G and H is the graph having vertex set $V(G) \cup V(H)$ and edge set $E(G) \cup E(H)$ and the join $G + H$ of G and H is the graph having vertex set $V(G) \cup V(H)$ and edge set $E(G) \cup E(H) \cup \{uv \mid u \in V(G), v \in V(H)\}$. A graph H is said to be obtained from a graph G by inserting a vertex $w (\notin V(G))$ into an edge uv of G if $V(H) = V(G) \cup \{w\}$ and $E(H) = (E(G) - \{uv\}) \cup \{uw, wv\}$. Two graphs H_1 and H_2 are said to be homeomorphic if both of them can be obtained from the same graph G by inserting vertices into the edges of G. The complement $\bar{G}$ of a graph G is the graph having vertex set V(G) such that two vertices in $\bar{G}$ are adjacent if and only if they are not adjacent in G. The line graph L(G) of a graph G is the graph having vertex set E(G) such that two vertices in L(G) are adjacent if and only if their corresponding edges in G are adjacent.

A connected graph is a graph such that any two vertices are connected by a path. A graph G which is not connected is the (disjoint) union of some connected subgraphs which are called the components of G. A component of a graph is odd if it has an odd number of vertices. The number of odd-components of G is denoted by o(G). A vertex v of G is a cut-vertex if $G - v$ has more components than that of G. Analogous to the cut-vertex is the concept of a bridge. A bridge of a graph G is an edge e such that the graph $G - e$ obtained from G by deleting the edge e has more components than that of G.

A (proper) vertex-colouring of G is a map $\pi : V(G) \to \{1,2,...\}$ such that no two adjacent vertices have the same image. The chromatic number $\chi(G)$ of G is the minimum cardinality of all possible images of vertex-colourings of G.

The following two theorems will be applied :

Dirac's theorem *If G is a graph of order $n \geq 3$ such that $\delta(G) \geq n/2$, then G is Hamiltonian.*

Tutte's theorem A graph G has a 1-factor if and only if

$$o(G - S) \leq |S| \quad \text{for all } S \subseteq V(G).$$

2. Groups acting on sets

In this section we shall define some basic graph-theorectic terms and state some theorems on group theory that will be used in this book.

Suppose X is a nonempty set with (or without) a structure. Then the set of all structure-preserving permutations of the elements of X forms a group under composition of maps. For instance, if G is a graph and X is the vertex set of G, then the set of all permutations of X preserving the adjacency of vertices forms a group, called the automophism group of G.

Historically, the theory of groups dealt at first with such permutation groups and later dealt with only abstract groups. However, it has been found that the notion of group actions (or groups acting) on sets, which passes an abstract group to a concrete permutation group, provides good counting techniques. As a result, the notion of group actions on sets plays an important role in the theory of finite groups.

We say that a group (an abstract group) G acts on a nonempty set X if to each g in G and each x in X there corresponds a unique element g(x) in X such that for every $x \in X$ and for every $g, h \in G$, $gh(x) = g(h(x))$ and $1(x) = x$, where 1 is the identity element in G.

Now suppose G acts on a set $X \neq \phi$. Then to each g in G, there corresponds a permutation ϕ_g in Σ_X, the set of all permutations of X, given by $\phi_g : x \to g(x)$. It is clear that $\phi : G \to \Sigma_X$ given by $\phi : g \to \phi_g$ is a (group) homomorphism. We call ϕ the permutation representation of G corresponding to the group action.

Conversely, suppose $\phi : G \to \Sigma_X$ is a homomorphism. Then G acts on X when we define $g(x) = \phi(g)(x)$ for each $g \in G$ and each $x \in X$. Thus a group action of G on X can be defined alternatively as a homomorphism from G to Σ_X.

From the second definition, we can see that the notion of a group acting on a set $X \neq \phi$ is more general than that of a permutation group on X, because in the former case unequal group elements can give rise to equal permutations, i.e. the map $\phi : g \to \phi_g$ need not be one-to-one. If the map $\phi : g \to \phi_g$ is one-to-one, then G is said to act faithfully on X.

Suppose G acts on a set $X \neq \phi$. It is not difficult to show that if we define a relation $\sim$ on X by setting $x_1 \sim x_2$ if there exists g in G such that $g(x_1) = x_2$, then $\sim$ is an equivalence relation on X. Hence, for each x in X, we can define the G-orbit of x, denoted by Orb(x), to be the set $\{g(x) \mid g \in G\}$ and the stabilizer G_x (or Stab(x)) of x in G to be the set $\{g \in G \mid g(x) = x\}$. It is not difficult to show that G_x is a subgroup of G and that $|\mathrm{Orb}(x)| = [G : G_x]$ where $[G : G_x]$ is the index of G_x in G. Thus if G is finite, then $|\mathrm{Orb}(x)| = |G|/|G_x|$.

Let G act on a set $X \neq \phi$. The action is said to be transitive if it has just one orbit; otherwise it is intransitive. An action of G on X is doubly transitive if for any two ordered pairs (x_1,x_2), (y_1,y_2) of distinct elements of X, there is some g in G such that $g(x_i) = y_i$, $i = 1, 2$. An action of G on X is said to be regular if it is transitive and $G_x = \{1\}$ for each x in X. Hence a regular action is faithful. The following theorem will be used in the study of vertex-transitive graphs in Chapter 3.

Theorem 2.1 If a finite group G acts transitively on X, then for any $x \in X$, $|X| = |G|/|G_x|$.

If G acts on X and x, $y \in X$ are such that $g(x) = y$, then it is not difficult to show that $G_y = g^{-1}G_xg$. This fact can be used to prove the following theorem.

Theorem 2.2 (Burnside's counting theorem) If a finite group G acts on $X \neq \phi$, then the number of orbits of G is

$$\frac{1}{|G|} \sum_{g\in G} \psi(g)$$

where $\psi(g) = |\{x \in X \mid g(x) = x\}|$.

We can define an action of G on itself by conjugation : for each g, $x \in G$, we write $x^g = g^{-1} xg$. Then $\mathrm{Orb}(x) = \{g^{-1}xg \mid g \in G\}$ is the conjugacy class of x in G and $\mathrm{Stab}(x) = \{g \in G \mid g^{-1} xg = x\} = \{g \in G \mid$

$xg = gx\} = C_G(x)$ is the centralizer of x in G. Hence, if G is a finite group having k distinct conjugacy classes, then from the fact that $|Orb(x)| = [G : G_x]$, we have the class equation of G :

$$|G| = \sum_{i=1}^{k} |Orb(x_i)| = \sum_{i=1}^{k} [G : C_G(x_i)] \tag{1}$$

where $x_1, \ldots, x_k$ are the representatives of the k conjugacy classes.

Let Z(G) be the set of elements x in G such that $C_G(x) = G$. Then Z(G) is the centre of G, and from (1) we have

$$|G| = |Z(G)| + \sum [G : C_G(y_i)] \tag{2}$$

where y_i runs through a set of representatives of the conjugacy classes which contain more than one element.

Suppose G is a finite group of order $p^k m$ where p is a prime and $p \nmid m$ (p does not divide m). Then a subgroup H of G such that $|H| = p^k$ is called a Sylow p-subgroup of G.

Using (2), H. Wielandt produced a very short proof of the following theorem which we shall apply in Chapter 3.

Theorem 2.3 (Sylow's theorem) Suppose G is a finite group of order $p^k m$ where p is a prime and $p \nmid m$. Then

(i) G contains a subgroup of order p^i for every $i \leq k$.

(ii) Any two Sylow subgroups of G are conjugate in G, i.e. if H_1 and H_2 are Sylow p-subgroups, then there exists $g \in G$ such that $H_2 = g^{-1} H_1 g$.

(iii) The number of Sylow p-subgroups of G is a divisor of m and is congruent to 1 modulo p.

(iv) Any subgroup of order p^i, $i \leq k$, is contained in a Sylow p-subgroup.

(For a 2-page proof of this theorem, see N. Jacobson: Basic Algebra 1, pp. 78-79.)

Suppose G acts transitively on X. For each subset Y of X and each g in G, let $g(Y) = \{g(y) \mid y \in Y\}$. A subset Y of X is said to be a block for the action if for each g in G, either $g(Y) = Y$ or $g(Y) \cap Y = \phi$. It is clear that ϕ, X and all the 1-element subsets of X are blocks for the action. These blocks are called the trivial blocks. The action is said to be primitive if the only blocks are the trivial blocks; otherwise the action is imprimitive.

For results concerning primitive group actions, the readers can refer to N. L. Biggs and A. T. White : Permutation Groups and Combinatorial Structures, London Mathematical Society Lecture Note Series 33, 1979; or J. S. Rose : A Course on Group Theory, Cambridge University Press, 1978.

We shall apply Polya's Pattern Counting Theorem in Chapter 3. Before we state Polya's theorem, we first define some terms.

Let D and R be nonempty sets. Let ψ be a map with domain D and range R and let G be a permutation group on D. We define a binary relation $\sim$ on the set R^D of all maps from D to R as follows : $\psi_1 \sim \psi_2$ if there exists $g \in G$ such that

$$\psi_1(d) = \psi_2(g(d)) \quad \text{for every } d \in D \tag{3}$$

It can be shown that this binary relation $\sim$ is an equivalence relation on R^D. The equivalence classes determined by $\sim$ are called the patterns. The patterns correspond to the distinct ways of distributing $|R|$ objects into $|D|$ cells when equivalence between ways of distribution is introduced by the group acting on D.

To each element r in R (called the store), we assign a weight w(r) which is an element in a commutative ring. The inventory of R is defined to be $\Sigma_{r \in R}\, w(r)$. Now for each $\psi \in R^D$, we define the weight $W(\psi)$ of ψ to be $\Pi_{d \in D}\, w(\psi(d))$, and for each $S \subseteq R^D$ we define the inventory of S to be $\Sigma_{\psi \in S}\, W(\psi)$. From the definitions, it follows that if $\psi_1 \sim \psi_2$, then $W(\psi_1) = W(\psi_2)$. Hence we can define the weight of a pattern P to be the weight $W(\psi)$ where $\psi \in P$.

Next, since each permutation $\psi(\in G)$ on $D = \{1,2,\ldots,n\}$ can be expressed uniquely as a product of disjoint cycles, for each $k = 1, 2, \ldots, n$, we let $j_k(\psi)$ to be the number of cycles of length k in the disjoint decomposition of ψ. The cycle index of G is defined by

$$Z(G ; x_1, \ldots, x_n) = \frac{1}{|G|} \sum_{\psi \varepsilon G} x_1^{j_1(\psi)} x_2^{j_2(\psi)} \ldots x_n^{j_n(\psi)}$$

where $x_1, \ldots, x_n$ are variables.

Using Burnside's counting theorem, we can prove

Theorem 2.4 (Polya's Theorem on Pattern Counting) Let D and R be nonempty sets and let G be a permutation group on D. Suppose the weights $w(r)$ of an element r in R and $W(P)$ of a pattern P are given as above. Then the inventory of the patterns of R^D is

$$Z(G ; \sum_{r\varepsilon R} w(r), \sum_{r\varepsilon R} [w(r)]^2, \ldots)$$

where $Z(G ; x_1, x_2, \ldots)$ is the cycle index of G.

Corollary 2.5 If all the weights are chosen to be equal to unity, then the number of patterns of R^D is $Z(G; |R|, |R|, \ldots, |R|)$.

A proof of Polya's Pattern Counting Theorem can be found in many textbooks on combinatorics, for instance, in B. Bollobás : Graph Theory-An Introductory Course, Graduate Text in Mathematics 63, Springer-Verlag, 1979; and in C. L. Liu : Introduction to Combinatorial Mathematics, McGraw-Hill, 1968.

2. EDGE-COLOURINGS OF GRAPHS

1. Introduction and definitions

The notion of an edge-colouring of a graph can be traced back to 1880 when Tait tried to prove the Four Colour Conjecture. (A detailed account of this can be found in many existing text books on Graph Theory and therefore we shall not repeat it here.) However, there was not much development during the period 1881-1963. A breakthrough came in 1964 when Vizing proved that every graph G having maximum valency Δ can be properly edge-coloured with at most $\Delta + 1$ colours ("proper" means that no two adjacent edges of G receive the same colour). This result generalizes an earlier statement of Johnson [63] that the edges of every cubic graph can be properly coloured with four colours.

Many of the results of this chapter will be concerned with the so-called 'critical graphs' introduced by Vizing in the study of classifying which graphs G are such that $\chi'(G) = \Delta(G) + 1$. The main reference of this chapter is Fiorini and Wilson [77].

We now give a few definitions. Let G be a graph or multigraph. A (proper, edge-) colouring π of G is a map $\pi : E(G) \to \{1,2,...\}$ such that no two adjacent edges of G have the same image. The chromatic index $\chi'(G)$ of G is the minimum cardinality of all possible images of colourings of G. Hence, if $\Delta = \Delta(G)$, then it is clear that $\chi'(G) \geq \Delta$ and Vizing's theorem says that $\Delta \leq \chi'(G) \leq \Delta + 1$. If $\chi'(G) = \Delta$, G is said to be of class 1, otherwise G is said to be of class 2. If π is a colouring of G such that the image set has cardinality k, then π is called a k-colouring of G. If $\chi'(G) \leq k$, then G is said to be k-colourable. Suppose π is a k-colouring of G having image set $\{1,2,...,k\}$. Let $C_\pi(v)$ or simply $C(v)$ be the set of colours used to colour the edges incident with v and let $C'_\pi(v)$ or simply $C'(v)$ be the set $\{1,2,...,k\} - C_\pi(v)$. If $i \in C(v)$ we say that colour i is present at v. If $j \in C'(v)$, we say that colour j is absent at v.

If π is a k-colouring of G, then π decomposes E(G) into a disjoint union of colour classes $E_1,\ldots,E_k$ in such a way that for each $e \in E_i$, $\pi(e) = i$. Hence, for $i \neq j$, each connected component of $E_i \cup E_j$ is either a cycle or a chain (open chain). If $j \in C(v)$ and $i \notin C(v)$, then the connected component of $E_i \cup E_j$ containing v is said to be a $(j,i)_\pi$-chain having origin v. From the definition, each E_i is a matching in G and thus certain results on the theory of matchings can be applied to the study of edge-colourings.

We now give a brief summary of the main results of this chapter.

In §1, we prove König's theorem which says that if G is a bipartite graph or multigraph, then $\chi'(G) = \Delta(G)$. We also prove that $\chi'(K_n) = n$ if n is odd and $\chi'(K_n) = n - 1$ if n is even. These two basic results are used to determine $\chi'(O_r^t)$ for the complete t-partite graph O_r^t in §2, which in turn is used to construct a class of chromatic index critical graphs in §4.

In §2, we give a generalization of Vizing's theorem due to Andersen and Gol'dberg (Theorem 2.2). From Theorem 2.2, we deduce Vizing's theorem and some results of Ore and Shannon. We also prove several sufficient conditions for a graph G to be of class 2.

In §3, we introduce the notion of (chromatic index) critical graphs which is the main tool for classifying which graphs are of class 2. We then give several properties of critical graphs. The main result of this section is Vizing's Adjacency Lemma, which is abbreviated as VAL. The results of this section are used very often in the subsequent sections of this chapter.

In §4, we produce several methods for constructing critical graphs. The most important method is the so-called HJ-construction due to Hajós and Jakobsen. We also produce several counter-examples to the Critical Graph Conjecture (which claims that every critical graph is of odd order).

In §5, we give some lower and upper bounds on the size of critical graphs. The main tool is Fiorini's inequality (Theorem 5.3). Applying Fiorini's inequality, Vizing's conjecture on the lower bound for the size of critical graphs G is verified for $\Delta(G) \leq 4$ and is shown to be "nearly true" for $\Delta(G) = 5$ and $\Delta(G) = 6$.

In §6, we first prove several general results on the minimum valency of a critical graph and then use them to construct all critical graphs of order at most 7. These results are applied in the study of 1-factorizations of regular graphs in §8.

In §7, we apply Fiorini's inequality and VAL to prove that every planar graph whose maximum valency is at least eight is necessarily of class 1. We also suggest an approach to prove the Planar Graph Conjecture (due to Vizing) which says that if G is a planar graph whose maximum valency is at least six, then G is of class 1.

In §8, we prove that if a regular graph G of order 2n has degree deg G = 2n − 3 or 2n − 4 such that $\deg G > 2[\frac{n+1}{2}] - 1$, then G is of class 1, i.e. G is 1-factorizable. Further results towards the resolution of the 1-Factorization Conjecture which says that if a regular graph G of order 2n has degree $\deg G > 2[\frac{n+1}{1}] - 1$, then G is 1-factorizable, are given in the exercises of this section.

In §9 and §10, we give some nontrivial applications of the theory of edge-colourings to vertex-colourings and to the reconstruction of latin squares.

In §11, we briefly mention some other interesting and important results on edge-colourings of graphs which we have not been able to discuss in detail due to lack of space.

The following theorem was proved by König in 1916 in connection with the factorization of graphs.

Theorem 1.1 (König [16]) *If* G *is a bipartite graph or multigraph having maximum valency* Δ, *then* $\chi'(G) = \Delta$.

Proof. We prove this theorem by induction on the size of G.

Let $e = vw \in E(G)$. By the induction hypothesis, $G - e$ has a Δ-colouring π. Suppose $C'(v) \cap C'(w) \neq \phi$. Let $i \in C'(v) \cap C'(w)$. Then we can extend π to a Δ-colouring of G by assigning colour i to the edge e. Hence we assume that $C'(v) \cap C'(w) = \phi$.

Let $i \in C'(v)$ and let $j \in C'(w)$. Then $j \in C(v)$ and $i \in C(w)$ and the $(j,i)_\pi$-chain C having origin v cannot have w as its terminus, otherwise $C \cup \{e\}$ forms an odd cycle in G, contradicting the fact that G is bipartite. Now, after interchanging the colours j and i in C, and assigning colour j to the edge e, we obtain a Δ-colouring of G. //

The chromatic index of K_n has been determined by several people using a variety of methods (see, for instance, Vizing [65a] and Behzad, Chartrand and Cooper [67]). The following proof is constructive and can be found in Berge [73] and Fiorini and Wilson [77].

Theorem 1.2 _For_ $n \geq 2$,

$$\chi'(K_n) = \begin{cases} n & \text{if } n \text{ is odd} \\ n-1 & \text{if } n \text{ is even} \end{cases}$$

Proof. We note that $\chi'(K_n) \geq \Delta(K_n) = n - 1$. Hence, for odd n, if we can show that $\chi'(K_n) \neq n - 1$ and establish an n-colouring of K_n, then $\chi'(K_n) = n$; and for even n, to prove that $\chi'(K_n) = n - 1$, we need only to establish an (n-1)-colouring of K_n.

First, suppose n is odd. If π is a colouring of G, then each colour class of π has cardinality at most $\frac{1}{2}(n - 1)$. Hence $e(G) \leq \frac{1}{2}(n - 1)\chi'(K_n)$, from which it follows that $\chi'(K_n) \geq n$. We now establish an n-colouring of K_n. We draw K_n as a regular n-gon and let $V(K_n) = \{v_1, v_2, \ldots, v_n\}$. Then the edges v_1v_n, v_2v_{n-1}, v_3v_{n-2}, ... are parallel. The edges v_2v_1, v_3v_n, v_4v_{n-1}, ... are also parallel and so on. We colour the first set of $\frac{1}{2}(n - 1)$ parallel edges by colour 1 and colour the second set of $\frac{1}{2}(n - 1)$ parallel edges by colour 2 and so on. This provides an n-colouring of K_n. Fig.2.1 illustrates partially the

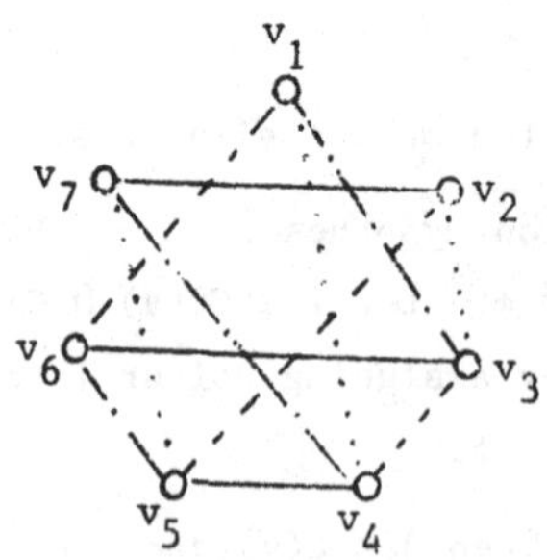

Figure 2.1

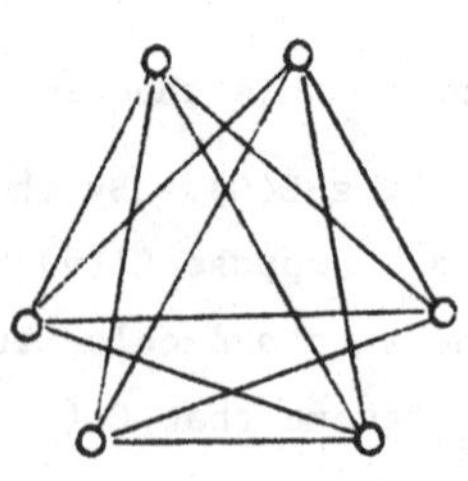
Figure 2.2

Next, suppose n is even. It is clear that $\chi'(K_2) = 1$. Hence, we assume that $n \geq 4$. Let $v \in V(K_n)$ and colour the graph $K_n - v$ in the manner described above. We observe that there is exactly one colour absent at each vertex in the above colouring of $K_n - v$ and that the colours absent at any two vertices of $K_n - v$ are pairwise distinct. Hence if colour i is absent at vertex u in $K_n - v$, then we colour the edge uv by colour i. This provides an (n-1)-colouring of K_n. //

Exercise 2.1

1^-. Prove that the chromatic index of a cubic Hamiltonian graph is 3.

2. Let P be the Petersen graph. Show that for any $v \in V(P)$, $\chi'(P - v) \neq 3$ and that $\chi'(P) = 4$.

   (Note that Castagna and Prins [72] have proved that every generalized Petersen graph G other than the Petersen graph P is of class 1.)

3. Prove or disprove that if G is a graph of class 1, then any induced subgraph H of G is also of class 1.

4. Find the chromatic index of each of the five platonic graphs.

5^+. Let C(r;t) be the graph obtained by arranging t copies of the null graph O_r into a cycle, and joining two vertices belonging to two different O_r's if and only if the two vertices lie in neighbouring members of the cycle (The graph C(2;3) is shown in Fig.2.2). Prove that C(r;t) is of class 2 if both r and t are odd, and is of class 1 otherwise (Parker [73]).

6. Prove that the chromatic index of the Coxeter graph (see Fig.3.8) is 4 (Biggs [73]).

7^+. Prove that if K_n has an (n-1)-colouring having the property that the group of permutations of the vertices which induce permutations on the colours acts triply transitively on the vertices, then $n = 2^d$ $(d \geq 2)$ or $n = 6$ and for each value of n the colouring is unique up to re-labelling of vertices and colours (Cameron [75a]).

8^+. Two edge-colourings ϕ and ψ of G are <u>isomorphic</u> if there is a permutation of the vertices of G which sends each member of the colour classes of ϕ into a member of the colour classes of ψ.

Prove that for any integer $n \geq 4$, there are two non-isomorphic colourings of K_{2n} (Wallis [73]).

9^+. Suppose π is a colouring of K_n, n even. Let $A(K_n;\pi)$ be the group of automorphisms of K_n which maps edges of the same colour to edges of the same colour and let $SA(K_n;\pi)$ be the subgroup of $A(K_n;\pi)$ that preserves the colour classes of π. Prove that (i) $|A(K_n;\pi)| \leq n^{1+\log_2 n}$; and (ii) $SA(K_n;\pi)$ is an elementary abelian 2-group and $|SA(K_n;\pi)|$ divides n (Cameron [75b]).

2. A generalization of Vizing's theorem

We shall first give a generalization of Vizing's theorem. This result is due to Andersen [77] and Gol'dberg [84]. (According to Gol'dberg [84], he published this result in Russian several years earlier.)

Let M be a multigraph without loops. An edge e of M is said to be <u>critical</u> if $\chi'(M - e) < \chi'(M)$. Suppose $\chi'(M) = q$, $e = xy$ is a critical edge of M, and π is a $(q-1)$-colouring of $M - e$. A <u>fan</u> $F = [xy_0, xy_1, \ldots, xy_n]$, where $y_0 = y$, $n \geq 1$, is a sequence of distinct edges $xy_0, xy_1, \ldots, xy_n$ such that for each $i \geq 1$, the edge xy_i is coloured with a colour absent at y_{i-1}. The set of all end-vertices different from x of all edges which are in at least one fan at x is denoted by A_x. We also define $B_x = N(x) - A_x$. For any two vertices u, v in M, let E(u,v) be the set of edges joining u and v, and let C(E(u,v)) be the set of colours colouring the edges of E(u,v).

Suppose $e = xy$ is a crtical edge of M with $\chi'(M) = q$, π is a $(q-1)$-colouring of $M - e$, and $F = [xy_0, xy_1, \ldots, xy_n]$ where $y_0 = y$, is a fan at x. We can now obtain another $(q-1)$-colouring of $M - e$ by uncolouring the edge xy_n and assigning colour $\pi(xy_i)$ to the edge xy_{i-1} for all $1 \leq i \leq n$. We shall refer to this process as <u>recolouring the fan</u> F.

Lemma 2.1 <u>Suppose</u> M <u>is a multigraph with a critical edge</u> $e = xy$, $\chi'(M) = q$, $d(x) \leq q - 1$, <u>and</u> π <u>is a</u> $(q-1)$-<u>colouring of</u> $M - e$. <u>Then</u>

(i) $C'(z) \cap C'(x) = \phi$ <u>for every</u> $z \in A_x$.

(ii) _If_ $j \in C'(x)$ _and_ $i \in C'(z)$, $z \in A_x$, _then the_ $(j,i)_\pi$_-chain having origin_ z _terminates at_ x.

(iii) _If_ $z, w \in A_x$ _are distinct, then_ $C'(z) \cap C'(w) = \phi$.

Proof. Suppose (i) is false. Let $y, y_1, \ldots, y_n$ where $z = y_n$ be the end-vertices of the edges of a fan F at x and let $i \in C'(z) \cap C'(x)$. Now we can obtain a (q-1)-colouring of M by recolouring the fan F and assigning colour i to the edge xz. This contradicts the fact that $\chi'(M) = q$.

Suppose (ii) is false and suppose $z \in A_x$ is such that the fan $F = [xy, xy_1, \ldots, xy_n]$, where $y_n = z$, has the minimum number of edges among all the fans for which (ii) fails. Then the $(j,i)_\pi$-chain C having origin z does not terminate at any of the vertices y_k, $k < n$, because n is assumed to be minimum. Therefore, after interchanging the colours in C, F remains a fan at x. But now $j \in C'(z) \cap C'(x)$ in the new (q-1)-colouring of M - e, contradicting (i).

Suppose (iii) is false. Let $i \in C'(z) \cap C'(w)$ and let $j \in C'(x)$. Then $i \neq j$ and, by (ii), both the $(j,i)_\pi$-chains having origins z and w terminate at x, which is impossible. //

Theorem 2.2 _For any finite multigraph_ $M = (V,E)$ _with maximum valency_ Δ,

$$\chi'(M) \leq \max\left\{\Delta, \sup_{x\in V} \sup_{\substack{A\subseteq N(x)\\ |A|=2}} \left[\frac{1}{2} \sum_{v\in A} (d(v) + \mu(v,x))\right]\right\}.$$

Proof. Suppose this theorem fails for a finite multigraph M'. By deleting some edges from M' we can obtain a multigraph M with $q = \chi'(M) = \chi'(M') > \Delta(M') \geq \Delta(M)$ such that M has a critical edge $e = xy$. Let π be a (q-1)-colouring of M - e. Then $d(x) = d_M(x) \leq q - 1$ and we can apply Lemma 2.1.

We note that from the definitions of A_x and B_x, and by Lemma 2.1(iii), we have

(1) for any vertex v in A_x and any vertex z in B_x, $C'(v) \cap C(E(z,x)) = \phi$.

(Suppose $j \in C'(v) \cap C(E(z,x))$ for some v in A_x and z in B_x. Let $v = y_p$ and let $F = [x, y_0, \ldots, y_{p-1}, y_p, y_{p+1}, \ldots]$ be a fan at x. Since $j \in C'(v) \cap C(E(z,x))$, the last vertex in F cannot be y_p. We recolour the fan $F' = [x, y_0, \ldots, y_{p+1}]$ and let this new colouring of $M - xy_{p+1}$ be ψ. Now $F'' = [x, y_{p+1}, y_p, \ldots]$ forms a fan in this new colouring ψ of $M - xy_{p+1}$ and the colour $\pi(xy_p) \in C'_\psi(y_p) \cap C'_\psi(y_{p+1})$, contradicting Lemma 2.1(iii).)

Hence, by Lemma 2.1 and (1), if $A_x = \{y, y_1, \ldots, y_n\}$ and $B_x = \{z_1, \ldots, z_m\}$, then

$$C'(x),\ C'(y),\ C'(y_1),\ \ldots,\ C'(y_n),\ C(E(z_1,x)),\ \ldots,\ C(E(z_m,x))$$

are pairwise disjoint. Now, since $\chi'(M - e) = q - 1$, we have

$$\begin{aligned} q - 1 &\geq |C'(x)| + |C'(y)| + \sum_{i=1}^{n} |C'(y_i)| + \sum_{j=1}^{m} |C(E(z_j,x))| \\ &= (q - 1 - (d(x) - 1)) + (q - 1 - (d(y) - 1)) + \sum_{i=1}^{n}(q - 1 - d(y_i)) \\ &\quad + (d(x) - \sum_{v \in A_x} \mu(v,x)). \end{aligned}$$

This gives $(n + 1)q \leq \sum_{v \in A_x} (d(v) + \mu(v,x)) + (n + 1) - 2$. Hence

$$\begin{aligned} q &\leq \left\lfloor \frac{\sum_{i=0}^{n}(d(y_i) + \mu(y_i,x))}{n + 1} + 1 - \frac{2}{n+1} \right\rfloor \\ &\leq \max_{\substack{A \subseteq N(x) \\ |A|=2}} \left[\frac{1}{2} \sum_{v \in A} (d(v) + \mu(v,x))\right], \end{aligned}$$

contradicting the assumption that the theorem is false. //

Corollary 2.3 (Vizing's theorem) *If* M *is a multigraph with maximum valency* Δ *and maximum multiplicity* μ, *then*

$$\chi'(M) \leq \Delta + \mu.$$

In particular, if G *is a graph with maximum valency* Δ, *then* $\chi'(G) = \Delta$ *or* $\chi'(G) = \Delta + 1$.

Corollary 2.4 (Ore [67]) *If* M *is a multigraph with maximum valency* Δ, *then*

$$\chi'(M) \leq \max\left\{\Delta,\ \sup\left[\frac{1}{2}(d(x) + d(y) + d(z))\right]\right\},$$

where the supremum is taken over all paths with three vertices x, y *and* z.

Corollary 2.5 (Shannon [49]) *If* M *is a multigraph with maximum valency* Δ, *then*

$$\chi'(M) \le [\tfrac{3}{2}\,\Delta].$$

Remarks.

(1) Bosák [72] proved that Corollaries 2.3 and 2.5 are also true for infinite multigraphs.

(2) Ehrenfeucht, Faber and Kierstead [84] gave an alternate proof of Theorem 2.2 using counting argument instead of fan recolouring.

The following theorem gives a sufficient condition for a graph to be of class 2. This theorem was proved by Beineke and Wilson [73], but it was implicit in the work of Vizing [65a].

Theorem 2.6 *Let* G *be a graph of order* n, *size* m *and having maximum valency* Δ. *If* $m > \Delta[\frac{n}{2}]$, *then* G *is of class* 2.

Proof. Suppose $\chi'(G) = \Delta$. Let π be a Δ-colouring of G and let $E_1,\ldots,E_\Delta$ be the colour classes. Then $|E_i| \le [\frac{n}{2}]$. Hence $m \le \Delta[\frac{n}{2}]$, yielding a contradiction. //

The *total deficiency* of a graph G is defined to be $\Sigma(\Delta(G) - d(v))$ where the summation extends to all vertices v of G.

Corollary 2.7 *If* G *is a graph of odd order, having maximum valency* Δ *and total deficiency less than* Δ, *then* G *is of class* 2.

Corollary 2.8 *If* G *is a regular graph of odd order, then* G *is of class* 2.

Corollary 2.9 *Suppose* H *is a graph of even order and is regular of degree* $d \ge 2$. *If* G *is obtained from* H *by inserting a new vertex into one edge of* H, *then* G *is of class* 2.

In §4, we shall apply the following theorem to construct a class of chromatic-index critical graphs.

Theorem 2.10 (Laskar and Hare [71]) Let $G = O_r^t$ be the complete t-partite graph having r vertices on each partition. Then

$$\chi'(G) = \begin{cases} r(t-1)+1 & \text{if } rt \text{ is odd} \\ r(t-1) & \text{if } rt \text{ is even} \end{cases}$$

Proof. Since O_r^t is regular of degree $r(t - 1)$, by Corollary 2.8, if rt is odd, then $\chi'(G) = r(t - 1) + 1$.

Suppose rt is even. To show that $\chi'(G) = r(t - 1)$, we need only to establish an $r(t-1)$-colouring of G.

Case 1. $r = 2s$.

Let $V(G) = (V_1 \cup V_2) \cup (V_3 \cup V_4) \cup \ldots \cup (V_{2t-1} \cup V_{2t})$ where each vertex of $V_i = \{v_{i1}, v_{i2}, \ldots, v_{is}\}$ is adjacent to every other vertex, except for those in V_i and the other V_j grouped in the parentheses with V_i.

From the fact that $\chi'(K_{2t}) = 2t - 1$, we know that the set of unordered pairs of numbers from the set $\{1,2,\ldots,2t-1,2t\}$ can be partitioned into $2t - 1$ sets $A_0, A_1, \ldots, A_{2t-2}$, where a given number occurs as a member of an unordered pair exactly once in each A_j. Without loss of generality, we may assume that

$$A_0 = \{\{1,2\}, \{3,4\}, \ldots, \{2t-1,2t\}\}.$$

Next, from the fact that $\chi'(K_{s,s}) = s$, we deduce that the set of ordered pairs of numbers from $\{1,2,\ldots,s\}$ can be partitioned into s sets $B_1, \ldots, B_s$ where in this case a given number occurs exactly once as a first component and exactly once as a second component of an ordered pair in each B_j. The sets $A_1, \ldots, A_{2t-2}$ and $B_1, \ldots, B_s$ will now be used to produce an $r(t-1)$-colouring of G.

Suppose v_{ip} and v_{jq} are two vertices that are adjacent in G. Then there is a unique g such that $\{i,j\} \in A_g$, and a unique h such that $(p,q) \in B_h$. Thus there exists a map $\pi : v_{ip}v_{jq} \to (g,h)$ from E(G) onto the set of ordered pairs (g,h), $1 \leq g \leq 2t - 2$, $1 \leq h \leq s$, such that adjacent edges have different images. This map π can be regarded as a proper colouring of G. Hence $\chi'(G) = (2t - 2)s = r(t - 1)$.

Case 2. t is even.

As in Case 1, let A_0, A_1, ..., A_{t-2} and $B_1, \ldots, B_r$ be determined from the values of $\chi'(K_t)$ and $\chi'(K_{r,r})$ respectively. Let $V = V_1 \cup V_2 \cup \ldots \cup V_t$, and let

$$V_i = \{v_{i1}, v_{i2}, \ldots, v_{ir}\}$$

where v_{ip} and v_{jq}, $i \neq j$, are adjacent vertices in G. Then there is a unique g such that $\{i,j\} \in A_g$ and a unique h such that $(p,q) \in B_h$. Thus there exists a map $\pi : v_{ip}v_{jq} \to (g,h)$ from E(G) onto the set of ordered pairs (g,h), $0 \leq g \leq t - 2$, $1 \leq h \leq r$, such that adjacent edges have different images. This map again can be regarded as a proper colouring of G. Hence $\chi'(G) = r(t - 1)$. //

Example $\chi'(O_4^3) = 8$.

Here

$A_1 = \{\{1,3\}, \{2,5\}, \{4,6\}\}$, $A_2 = \{\{1,4\}, \{2,6\}, \{3,5\}\}$,
$A_3 = \{\{1,5\}, \{2,4\}, \{3,6\}\}$, $A_4 = \{\{1,6\}, \{2,3\}, \{4,5\}\}$,
$B_1 = \{(1,1), (2,2)\}$, $B_2 = \{(1,2), (2,1)\}$.

Using the notation and the construction given in the proof of Case 1, the edges of the graph O_4^3 can be coloured with the eight colours (1,1), (1,2), (2,1), (2,2), (3,1), (3,2), (4,1) and (4,2) as shown in the following table.

Edges								
	$v_{11}v_{31}$	$v_{11}v_{32}$	$v_{11}v_{41}$	$v_{11}v_{42}$	$v_{11}v_{51}$	$v_{11}v_{52}$	$v_{11}v_{61}$	$v_{11}v_{62}$
	$v_{12}v_{32}$	$v_{12}v_{31}$	$v_{12}v_{42}$	$v_{12}v_{41}$	$v_{12}v_{52}$	$v_{12}v_{51}$	$v_{12}v_{62}$	$v_{12}v_{61}$
	$v_{21}v_{51}$	$v_{21}v_{52}$	$v_{21}v_{61}$	$v_{21}v_{62}$	$v_{21}v_{41}$	$v_{21}v_{42}$	$v_{21}v_{31}$	$v_{21}v_{32}$
	$v_{22}v_{52}$	$v_{22}v_{51}$	$v_{22}v_{62}$	$v_{22}v_{61}$	$v_{22}v_{42}$	$v_{22}v_{41}$	$v_{22}v_{32}$	$v_{22}v_{31}$
	$v_{41}v_{61}$	$v_{41}v_{62}$	$v_{31}v_{51}$	$v_{31}v_{52}$	$v_{31}v_{61}$	$v_{31}v_{62}$	$v_{41}v_{51}$	$v_{41}v_{52}$
	$v_{42}v_{62}$	$v_{42}v_{61}$	$v_{32}v_{52}$	$v_{32}v_{51}$	$v_{32}v_{62}$	$v_{32}v_{61}$	$v_{42}v_{52}$	$v_{42}v_{51}$
Images	(1,1)	(1,2)	(2,1)	(2,2)	(3,1)	(3,2)	(4,1)	(4,2)

Exercise 2.2

1. Applying Theorem 2.6, prove that if H is a graph of odd order such that H has a vertex of valency t and all other vertices are of valency $\Delta = \Delta(H)$, then any graph G obtained from H by deleting not more than $\frac{1}{2} t - 1$ edges is of class 2.

2. Let G be a graph with <u>edge-independence number</u> α, i.e. G has a maximum matching of cardinality α. Show that if $e(G) > \Delta(G)\alpha$, then G is of class 2.

3. Prove that if G is a regular graph which contains a cut-vertex, then G is of class 2.

4. Let G be a graph of order n. Prove that

   (a) if n is even, then
   $n - 1 \le \chi'(G) + \chi'(\bar{G}) \le 2n - 2$ and $0 \le \chi'(G)\chi'(\bar{G}) \le (n - 1)^2$;

   (b) if $n \ge 3$ is odd, then

   $n \le \chi'(G) + \chi'(\bar{G}) \le 2n - 3$ and $0 \le \chi'(G)\chi'(\bar{G}) \le (n - 1)(n - 2)$.

   Show that all these results are best possible (Vizing [65a], Alavi and Behzad [71]).

5. Following the proof of Theorem 2.10, find a 6-colouring of O_2^4.

6.[+] Prove that if G is an infinite graph such that every finite subgraph of G is k-colourable, then G is k-colourable (Neumann [54]).

7. A graph G is said to be <u>uniquely k-colourable</u> if any two k-colourings of G induce the same partititon of E(G). Show that every uniquely k-colourable graph $G \ne K_3$ has $\Delta(G) = k$. Hence, or otherwise, show that every uniquely k-colourable regular graph is Hamiltonian and that every uniquely 3-colourable cubic graph contains exactly three Hamilton cycles (Greenwell and Kronk [73]).

8. Prove that every connected regular graph of order 4 or 6 is of class 1.

3. Critical graphs

We have shown that bipartite graphs are of class 1, the graph O_r^t is of class 1 if and only if rt is even, the complete graph K_n is of class 1 if and only if n is even and that any regular graph of odd order is of class 2. Relatively speaking, there are considerably more class 1 graphs than class 2 graphs. This fact is reflected by a theorem of Erdös and Wilson [77] which says that the probability of a graph of order p, which is of class 1, approaches 1 as p tends to infinity. However, the problem of determining which graphs belong to which class is extremely difficult. One way of tackling this problem is through the study of critical graphs which was first introduced by Vizing.

A graph G is said to be (chromatic-index) <u>critical</u> if G is connected, of class 2, and $\chi'(G - e) < \chi'(G)$ for any edge e of G. If G is critical and $\Delta(G) = \Delta$, then G is said to be <u>Δ-critical</u>. This notion is parallel to the notion of criticality with respect to vertex-colourings which was introduced by Dirac in 1952.

The importance of this concept of criticality is that once we know that a graph G is Δ-critical, and if G^* is obtained from G by adding edges joining non-adjacent vertices of G such that $\Delta(G^*) = \Delta(G)$, then G^* is definitely of class 2. This certainly helps quite a lot in deciding which graphs are of class 2.

There are other notions of criticality with respect to edge-colourings. However, among these notions, the one introduced above seems to be the most important one in the sense that all other known notions of critical graphs are related to it (for details, see Hilton [77]).

The following two simple results were first observed by Vizing [65a].

Theorem 3.1 <u>Suppose</u> G <u>is</u> Δ-<u>critical and</u> $vw \in G$. <u>Then</u> $d(v) + d(w) \geq \Delta + 2$.

Proof. Let π be a Δ-colouring of $G - vw$. If $(d(v) - 1) + (d(w) - 1) < \Delta$, then $C'(v) \cap C'(w) \neq \phi$ and π can be extended to a Δ-colouring of G. //

Theorem 3.2 *A critical graph contains no cut-vertices.*

Proof. Suppose G is a Δ-critical graph containing a cut-vertex v. Let $H_1, \ldots, H_r$ be the components of G − v. Each of the subgraphs G_i, i = 1, ..., r obtained by joining v to H_i is Δ-colourable. These Δ-colourings of G_i can be combined to yield a Δ-colouring of G by making the colours of the edges incident with v all different. This contradiction shows that G cannot contain a cut-vertex. //

The next theorem shows that a class 2 graph G contains a whole range of k-critical subgraphs where k runs from 2 to $\Delta(G)$. This theorem will be used in the construction of critical graphs.

Theorem 3.3 (Vizing [65b]) *If G is a graph of class 2, then G contains a k-critical subgraph for each k satisfying $2 \leq k \leq \Delta(G)$.*

Proof. Let $\Delta(G) = \Delta$. We can obtain a Δ-critical subgraph H of G by removing all the edges of G whose deletion does not lower the chromatic index.

We now prove that G also contains a k-critical subgraph for each k satisfying $2 \leq k < \Delta$. It is clear that now $\Delta \geq 3$. Let $u \in V(H)$ be such that $d(u) = \Delta$. Since $\Delta \geq 3$, $|V(H)| \geq 4$ and thus H has two vertices v and w such that $v \neq u \neq w$ and $vw \in H$. Let π be a Δ-colouring of H − vw and let $E_1, \ldots, E_\Delta$ be the colour classes of π. Since $C'(v) \cap C'(w) = \phi$, there are $i \in C'(v)$ and $j \in C'(w)$. By taking away any $\Delta - k$ of the colour classes $E_1, \ldots, E_\Delta$ which are not E_i or E_j, we obtain a subgraph J of H such that $\Delta(J) = k$ and $\chi'(J) = k + 1$. (For each z ($\neq$ v, w) $\in V(H)$, after taking away $\Delta - k$ colour classes, the remaining edges incident with z are coloured with at most $\Delta - (\Delta - k) = k$ colours. Hence $d_J(z) \leq k$. For the vertices v and w, after taking away $\Delta - k$ colour classes, the remaining edges in H − vw which are incident with v or w are coloured with at most k − 1 colours. Hence $\Delta(J) \leq k$. However, it is clear that $d_J(u) = k$.) After removing some edges from J, if necessary, as in the first part of this proof, we obtain a k-critical subgraph of G. //

We now prove a generalization of another important result of Vizing about the structure of critical graphs.

Theorem 3.4 (Andersen [77]) Let $M = (V,E)$ be a multigraph for which

$$q = \chi'(M) = \max_{u \in V} (d(u) + \max_{v \in V} \mu(v,u))$$

and for which M has a critical edge $e = xy$. Then x is adjacent to at least $q - (d(y) + \mu(y,x)) + 1$ vertices v different from y such that $d(v) + \mu(v,x) = q$.

Proof. Let $t = d(y) + \mu(y,x)$. Consider a $(q-1)$-colouring π of $M - e$. We shall show that the $q - t + 1$ vertices mentioned in the statement of the theorem all belong to A_x. Using Lemma 2.1 (obviously $d(x) \leq q - 1$) we get (as in the proof of Theorem 2.2)

$$q - 1 \geq \sum_{v \in A_x - \{y\}} (q - 1 - d(v)) + (q - d(y)) + (q - d(x))$$

$$+ (d(x) - \sum_{v \in A_x} \mu(v,x)) \quad \text{giving}$$

$$\sum_{v \in A_x - \{y\}} (q - d(v) - \mu(v,x)) \leq (|A_x| - 1) - (q - t + 1) .$$

Now every term of the sum on the left-hand side of this inequality is non-negative. Hence there are at least $q - t + 1$ vertices v in $A_x - \{y\}$ for which $d(v) + \mu(v,x) = q$. //

Corollary 3.5 Let $M = (V,E)$ be a critical multigraph for which

$$q = \chi'(M) = \max_{u \in V} (d(u) + \max_{v \in V} \mu(v,u)) .$$

Put $m(M) = \min_{u \in V} (d(u) + \min_{v \in N(u)} \mu(v,u))$. Then we have

(i) every vertex x of M is adjacent to at least two vertices v for which $d(v) + \mu(v,x) = q$; and

(ii) M contains at least $\max\{3, q - m(M) + 1\}$ vertices w for which

$$d(w) + \max_{u \in V} \mu(u,w) = q .$$

Suppose G is a graph. If $v \in G$ is of valency $\Delta(G)$, then v is called a _major vertex_, otherwise a _minor vertex_. From the above theorem and its corollary, we have the so-called Vizing's Adjacency Lemma which is abbreviated as VAL, because it will be used very often throughout the rest of this chapter.

Corollary 3.6 (VAL) _Let_ G _be a_ Δ-_critical graph and let_ $vw \in G$ _where_ $d(v) = k$. _We have_

(i) _if_ $k < \Delta$, _then_ w _is adjacent to at least_ $\Delta - k + 1$ _major vertices of_ G;

(ii) _if_ $k = \Delta$, _then_ w _is adjacent to at least two major vertices of_ G;

(iii) G _has at least_ $\Delta - \delta(G) + 2$ _major vertices; and_

(iv) G _has at least three major vertices._

Exercise 2.3

1. Show that there does not exist a critical graph having exactly two minor vertices, one of which is of valency 2 (Yap [80]).

2. Prove that if G is a Δ-critical graph and if M is a nonzero matching in G, then

   (a) there exists a $(\Delta+1)$-colouring of G in which M is a colour class; and

   (b) $\chi'(G - M) = \chi'(G) - 1$.

3. Let G be a Δ-critical graph, $vw \in G$ and let π be a Δ-colouring of $G - vw$. Prove that

   (a) $|C(v) \cup C(w)| = \Delta$;

   (b) $|C(v) \cap C(w)| = d(v) + d(w) - \Delta - 2$;

   (c) $|C(v) \setminus C(w)| = \Delta + 1 - d(w)$; and

   (d) $|C(w) \setminus C(v)| = \Delta + 1 - d(v)$ (Berge [73; p.254]).

4. Verify that the following graph is 3-critical.

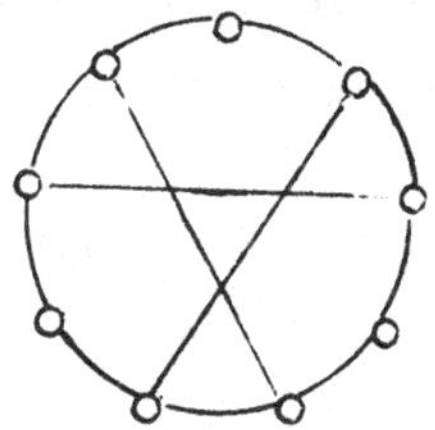

Figure 2.3

5^{+}. Verify that the graph given in Fig.2.4 is 3-critical (Gol'dberg [79b]).

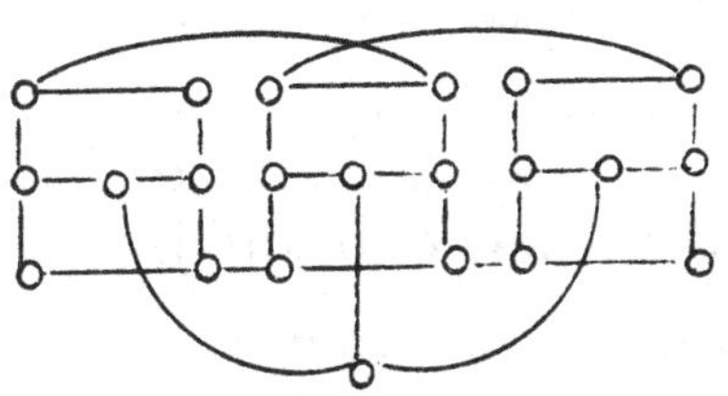

Figure 2.4

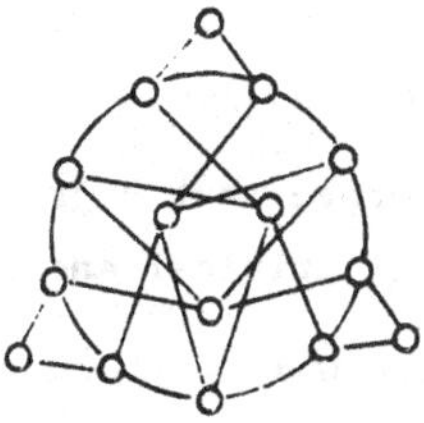

Figure 2.5

6^{+}. Verify that the graph given in Fig. 2.5 is 4-critical. (This graph was produced by A. G. Chetwynd, see Yap [80].)

7. Prove that there does not exist a 4-critical graph of even order n having valency-list $23^2 4^{n-3}$ (For definition of <u>valency-list</u>, see §5.) and that for $\Delta > 10$, there does not exist a Δ-critical graph of even order n having valency-list $23^2 \Delta^{n-3}$.

8*. (a) Does there exist a 4-critical graph of odd order n having valency-list $23^2 4^{n-3}$?

(b) For $\Delta = 6$ or 8, does there exist a Δ-critical graph of even order n having valency-list $23^2 \Delta^{n-3}$?

9^{-}. Prove that if G is a Δ-critical graph of odd order n having size at least $\Delta(\frac{n-3}{2}) + \delta(G) + 1$, then for every $x \in G$ such that $d(x) = \delta(G)$, $G - x$ has a 1-factor.

10*. Is it true that if G is any Δ-critical graph of odd order, then for each $x \in G$ such that $d(x) = \delta(G)$, $G - x$ has a 1-factor ?

11. Prove that if G is a Δ-critical graph of odd order n having size at least $\Delta(\frac{n-3}{2}) + 2$, then G has a minor vertex y such that G - y has a 1-factor.

12. Let H be the subgraph induced by the major vertices of a graph G. Prove that if H contains no cycles, then G is of class 1 (Fiorini and Wilson [77; p.78]).

13. A graph G is pc-critical (proper-class critical) if G is of class 2, but every proper subgraph G' of G is of class 1. A graph G is pi-critical (proper-index critical) if $\chi'(G') < \chi'(G)$ for every proper subgraph G' of G. A graph G is ei-critical (edge-index critical) if $\chi'(G - e) < \chi'(G)$ for each edge e of G.

    (a) Prove that a graph is pc-critical if and only if it is a cycle of odd length.

    (b) Prove that a graph is pi-critical if and only if it is ei-critical and has no isolated vertices (Hilton [77]).

14. Suppose G is a graph of order 2n having r major vertices. Prove that if $\Delta(G) > n + \frac{5}{2}r - 4$, then G is not critical (Chetwynd and Hilton [-b]).

4. Constructions for critical graphs

In this section we shall introduce various methods for constructing critical graphs. We need the following definitions. Suppose $x \in G$ is such that $d(x) = m > 2$, and $N(x) = \{x_1, \ldots, x_m\}$. We say that the graph H is obtained from G by splitting x into two vertices u and v ($u, v \notin G$) if

$$V(H) = V(G - x) \cup \{u, v\} \quad \text{and}$$

$$E(H) = E(G - x) \cup \{uv, ux_1, \ldots, ux_r, vx_{r+1}, \ldots, vx_m\}$$

for some r satisfying $1 \leq r < m$. From this definition, we see that if H is obtained from a graph G by inserting a vertex v into an edge whose end-vertices are both of valency at least 2, then H can be considered as a graph obtained from G by splitting a vertex x into two vertices x and v. Two graphs obtained from the graph $K_{4,4}$ by splitting a vertex into two vertices u and v are shown in Fig.2.6.

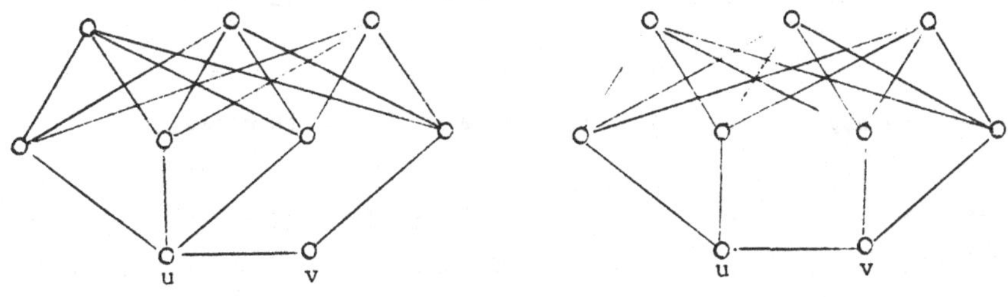

Figure 2.6

We first prove

Lemma 4.1 If G is a connected graph with forbidden induced subgraph $K_2 \cup 0_1$, then either $G = K_n$ or $G = 0_{r_1} + 0_{r_2} + \ldots + 0_{r_t} + K_s$ where $r_i \geq 2$ for all $i = 1, 2, \ldots, t$.

Proof. Suppose $G \neq K_n$. Let u and v be two non-adjacent vertices of G. By the 'forbidden induced subgraph condition', we have $N(u) = N(v)$. Let $M(u) = V(G) - N(u)$ and let $W = M(u) - \{u,v\}$. If $W = \phi$, then $G = 0_2 + H$ where $H = \langle N(u) \rangle$. If $W \neq \phi$, let $w \in W$. Again, by the 'forbidden induced subgraph condition', $N(w) = N(u)$ and $\langle u,v,w \rangle \cong 0_3$. Hence, in any case, $\langle M(u) \rangle \cong 0_{r_1}$ for some integer $r_1 \geq 2$.

Next, suppose $H \neq K_m$ or 0_m. We first show that H is connected. If H is not connected, let H_1 and H_2 be two components of H. Since $H \neq 0_m$, H_1 (say) has two adjacent vertices x and y. Now for any vertex z of H_2, $\langle x,y,z \rangle \cong K_2 \cup 0_1$ is a forbidden induced subgraph of G. This contradiction shows that H is connected and the lemma follows by induction on the order of G. //

Corollary 4.2 If a graph G is connected, regular of degree at least 2 with forbidden induced subgraphs $K_2 \cup 0_1$, then $G = 0_r^t$ for some integers $r \geq 1$, $t \geq 3$ or $r \geq 2$, $t \geq 2$.

From the above corollary and Theorem 2.10, we have

Lemma 4.3 If a graph G is connected, regular of degree at least 2 with forbidden induced subgraphs $K_2 \cup O_1$ and G is of class 1, then $G = O_r^t$ where rt is even.

We now apply Lemma 4.3 to prove the following theorem.

Theorem 4.4 (Yap [81a]) Let r and t be two positive integers such that $rt \geq 4$ is even. Then the graph H obtained from $G = O_r^t$ by splitting a vertex x into two vertices u and v is critical.

Proof. Let $\Delta = \Delta(G)$ and let $rt = n$. Then $\Delta = r(t - 1)$. If $\Delta = 2$, then $G = O_2^2 = K_{2,2}$ and $H = C_5$ is 2-critical. Hence, from now on, we assume that $\Delta \geq 3$.

For convenience, we introduce an auxiliary graph $K = H - uv$. We know that $|H| = |G| + 1 = n + 1$ is odd and that $e(H) = \frac{1}{2} n\Delta + 1 > \Delta[\frac{n+1}{2}]$. Hence, by Theorem 2.6, H is of class 2.

Let $J = G - x$ and let μ be a Δ-colouring of G. We now modify μ to a Δ-colouring π of K as follows : for any edge f of K,

$$\pi(f) = \begin{cases} \mu(f) & \text{if } f \in J \\ \mu(xy) & \text{if } f = uy \text{ or } vy,\ y \in N_G(x) \end{cases}$$

We observe that each of the colours 1, ..., Δ appears exactly once at either vertex u or vertex v in this Δ-colouring π of K.

We shall now prove that π can be modified to a Δ-colouring λ of $H - e$ for any edge e of H and so H is critical.

Suppose $e = uv$. Then $H - e = K$ and λ can be chosen as π.

Suppose $e = uy$, $y \in N_G(x)$. (The argument for the case $e = vz$, $z \in N_G(x)$ is similar.) We choose λ as follows : for any edge f of $H - e$,

$$\lambda(f) = \begin{cases} \pi(f) & \text{if } f \in K \\ \pi(uy) & \text{if } f = uv \end{cases}$$

Suppose $e = yz \in J$. We assume that $\pi(e) = 1$.

Case 1. uy, vz ε H.

Since $\Delta > 3$, without loss of generality, we may assume that $\pi(vw) = 1$ where w ε J. We also let $\pi(uy) = k$.

Let C be the $(1,k)_\pi$-chain having origin v. We note that besides y and z, u and v are the only minor vertices in K - e. Therefore, the terminus of C is either u, or y, or z. In this case, it is clear that the terminus of C must be z because colour 1 is absent at y and u in the Δ-colouring π of K - e and $\pi(uy) = k$. By interchanging the colours in C and assigning colour 1 to the edge uv, we get a Δ-colouring λ of H - e. (We call such an argument the Kempe-chain argument.)

Case 2. vy, vz ε H.

(The argument for the case uy, uz ε H is similar.)

If colour 1 is absent at v in the Δ-colouring π of K, we replace colour $\pi(vy)$, which has been assigned to vy, by colour 1 and we assign colour $\pi(vy)$ to the edge uv to get a Δ-colouring λ of H - e.

Suppose $1 \varepsilon C_\pi(v)$ and $i \notin C_\pi(v)$. Let C be the $(1,i)_\pi$-chain having origin v. If the terminus of C is u, then y, z $\notin$ C. By interchanging the colours in C, we reduce it to the previous case. Hence we may assume, without loss of generality, that the terminus of C is y. Now z, u $\notin$ C and by interchanging the colours 1 and i in C, we find that colour 1 is absent from both vertices u and v in this new colouring of H. Thus, by assigning colour 1 to the edge uv, we get a Δ-colouring λ of H - e.

Case 3. vy ε H, uz $\notin$ H.

(The argument for the case uy ε H, vz $\notin$ H is similar.)

The proof for this case is exactly the same as the proof for Case 2.

Cases 1 and 2 correspond to the case $\langle x,y,z\rangle \cong K_3$ and Case 3 corresponds to the case $\langle x,y,z\rangle \cong P_3$. The proof is now complete. //

Corollary 4.5 For any two integers Δ and σ such that $\Delta > 3$ and $2 < \sigma < \Delta$, we can construct a Δ-critical graph containing a vertex of valency σ.

Remarks

1. Fiorini (see Fiorini and Wilson [77;p.70]) has shown that if the graph H is obtained from O_m^2, K_{2m}, O_2^m by inserting a vertex into an edge of these graphs, then H is critical. Hence Fiorini's result is a special case of a subclass of the class of critical graphs constructed in Theorem 4.4.

2. The implicit 'forbidden induced subgraph condition' in Theorem 4.4 is not redundant as can be seen from the graph G given in Fig.2.7 : G is 3-regular and is of class 1. But $K_2 \cup O_1$ is an induced subgraph of G. The graph H - pq is the Petersen graph minus a vertex and is of class 2 (see Ex.2.1(2)). In fact, H - pq is 3-critical (see Ex.2.3(4)).

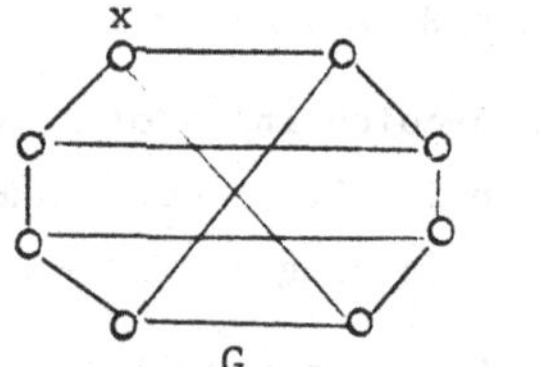

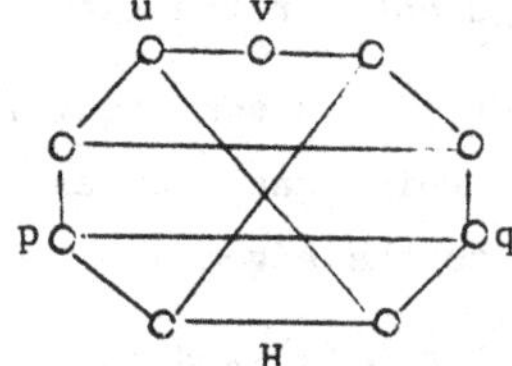

Figure 2.7

3. Chetwynd and Hilton [84] proved that if G is obtained by removing any (n-3)/2 edges from K_n, for odd $n \geq 5$, then G is (n-1)-critical. Plantholt [-a] proves that any graph G with $\Delta(G) = n$ which can be obtained by removing $2n - 1$ edges from $K_1 + K_{n,n}$ is n-critical. Both of these results generalize a special case of Theorem 4.4. It is nice to know whether any graph G with $\Delta(G) = r(t - 1)$ obtained by removing any $rt - 1$ edges from $K_1 + O_r^t$, where $rt \geq 4$ is even, is r(t-1)-critical.

The next theorem provides a construction for Δ-critical graphs from some known Δ-critical graphs of smaller order. This construction is due to Jakobsen [73] but the method was originally due to Hajós [61]. We shall call this construction the HJ-construction.

Theorem 4.6 (HJ-construction) Let G and H be two Δ-critical graphs and K be a graph obtained from G and H by identifying $u \in V(G)$ and $v \in V(H)$ such that $d_G(u) + d_H(v) \leq \Delta + 2$, removing edges $uz \in G$ and $vz' \in H$ and joining the vertices z and z'. Then K is also Δ-critical.

Proof. It is clear that K is connected and $\Delta(K) = \Delta$.

We first prove that K is of class 2. Suppose otherwise. Let π be a Δ-colouring of K and let $\pi(zz') = 1$. Then there is an edge e of G incident with u such that $\pi(e) = 1$ otherwise $\chi'(G) = \Delta$. Similarly, there is an edge e' of H incident with v such that $\pi(e') = 1$. But since u and v have been identified, e and e' are adjacent in K and thus K has two adjacent edges receiving the same colour under π, which is false (see Fig.2.8).

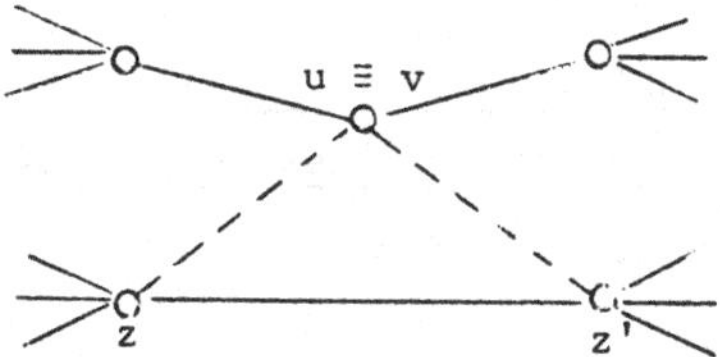

Figure 2.8

It remains to show that $\chi'(K - e) = \Delta$ for any edge e of K. If $e = zz'$, this is clear, since any two Δ-colourings of $G - uz$ and $H - vz'$ can be combined to give a Δ-colouring of $K - zz'$. We may now assume that $e \in G - uz$ (the case that $e \in H - vz'$ can be settled in a similar way). It is clear that $\chi'(H - vz' + zz') = \Delta$ and for every Δ-colouring ϕ of $H - vz' + zz'$, there is an edge vy in H such that $\phi(vy) = \phi(zz')$. It is also clear that any Δ-colouring ψ of $G - e$ gives rise to a Δ-colouring ψ' of $G - uz - e + zz'$ in which the colour assigned to zz' is absent at u. Finally, the two Δ-colourings ϕ of $H - vz' + zz'$ and ψ' of $G - uz - e + zz'$ can be combined to give a Δ-colouring of $K - e$. //

The critical graphs constructed by Theorem 4.4 are of odd order. If the critical graphs G and H given in Theorem 4.6 are of odd order, then the critical graph K constructed from the HJ-construction is also of odd order. Our next theorem provides a construction for a Δ-critical graph H of even order from a Δ-critical graph G of odd order if G has a vertex u of valency 2 and another vertex $v \neq u$ of valency at most $(\Delta+2)/2$.

Lemma 4.7 Let $\Delta \geq 3$ be an integer and let G be a Δ-critical graph having two vertices u and v each of valency 2. Suppose $N(u) = \{w,x\}$ and

$N(v) = \{y,z\}$. *Then*

(i) *there exists a* Δ*-colouring* π *of* $G - vz$ *such that* $\pi(vy) = 1$, $\pi(ux) = 1$ *and* $\pi(uw) = 2$; *and*

(ii) *there exists a* Δ*-colouring* λ *of* $G - vz$ *such that* $\lambda(vy) = 1$, $\lambda(ux) = 2$ *and* $\lambda(uw) = 3$.

A proof of this lemma can be found in Yap [80]. Since the proof is easy, we shall leave it as an exercise.

Lemma 4.8 *Let* $\Delta \geq 4$ *be an integer. If there exists a critical graph* G *of odd order such that* G *has two vertices of valency* 2, *then there exists a* Δ*-critical graph of even order.*

Proof. Let u and v be two vertices of G of valency 2, $N(u) = \{w,x\}$ and $N(v) = \{y,z\}$. Let G' be a copy of G such that each of u', v', w', x', y', z',... in G' corresponds respectively to u, v, w, x, y, z, ... in G.

Let H_1 be the Δ-critical graph constructed, using the HJ-construction, from G and G' by identifying the vertices v and v', removing the edges vz and $v'z'$ and joining the vertices z and z'.

Let H be the graph obtained from H_1 by identifying u and u'. Then $d_H(u) = 4$, H is of even order and is of class 2. We now prove that $\chi'(H - e) = \Delta$ for any edge e in H.

We may assume that e is either zz' or an edge in $G - vz$ since the corresponding argument for $e \in G' - v'z'$ is exactly the same.

Suppose $e = uw$. (The proof for the case $e = ux$ is similar.) By Lemma 4.7, there exists a Δ-colouring π of $G - e$ such that $\pi(ux) = 2$, $\pi(vy) = 2$ and $\pi(vz) = 1$. Again, by Lemma 4.7, there exists a Δ-colouring λ of $G' - v'z'$ such that $\lambda(v'y') = 1$, $\lambda(u'w') = 3$ and $\lambda(u'x') = 1$. Since G' is critical, colour 1 is absent at z'. Hence the Δ-colouring π of $G - uw + zz'$ and the Δ-colouring λ of $G' - v'z' + zz'$ can be combined to give a Δ-colouring of $H_1 - e$ as shown in Fig.2.9. Thus $H - e$ is Δ-colourable.

Suppose $e = zz'$. Again, by Lemma 4.7, there exists a Δ-colouring π of $G - vz$ such that $\pi(vy) = 2$, $\pi(uw) = 4$, $\pi(ux) = 2$ and there exists a

Δ-colouring λ of $G' - v'z'$ such that $\lambda(v'y') = 1$, $\lambda(u'w') = 3$, $\lambda(u'x') = 1$. Then π and λ can be combined to give a Δ-colouring $H_1 - zz'$ and so $H - zz'$ is also Δ-colourable.

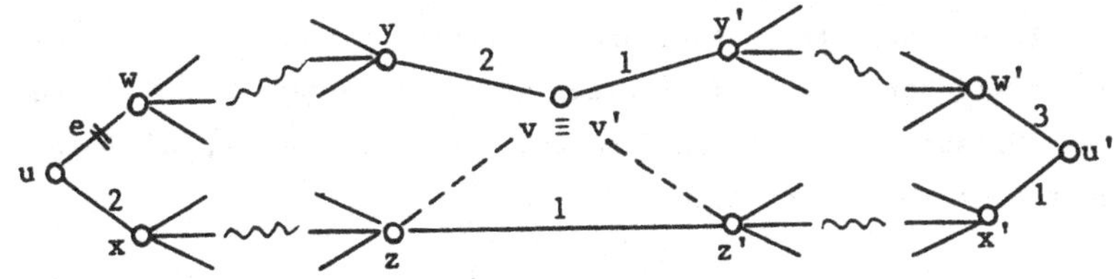

Figure 2.9

Suppose e is not incident with u or v. Let π be a Δ-colouring of $H_1 - e$ such that $\pi(zz') = 1$, $\pi(vy) = 2$, $\pi(uw) = j$ and $\pi(ux) = k$.

If $\{j,k\} \cap \{1,2\} = \phi$, then by Lemma 4.7 we can assign the colours 1 and 2 to the edges u'w', u'x' under π and so $H - e$ is Δ-colourable.

If $\{j,k\} = \{1,2\}$, then by Lemma 4.7 we can assign the colours 3 and 4 to the edges u'w', u'x' under π and so $H - e$ is Δ-colourable.

If $j = 2$, $k \neq 1$, then by Lemma 4.7 we can assign the colours 1 and $i \neq 1,2,k$ to the edges u'w', u'x' under π and so $H - e$ is Δ-colourable.

If $j = 1$, $k \neq 2$, then by Lemma 4.7 we can assign any two colours other than colours 1 and k to the edges u'w', u'x' under π and so $H - e$ is Δ-colourable.

Finally, the case $e = vy$ can be settled in a similar way. //

Theorem 4.9 (Yap [80]) Let $\Delta \geq 4$ be an integer. If there exists a Δ-critical graph K of odd order such that K has a vertex u of valency 2 and another vertex $v \neq u$ of valency at most $(\Delta+2)/2$, then there exists a Δ-critical graph of even order.

Proof. Let K' be a copy of K such that each of u', v', ... in K' corresponds respectively to u, v, ... in K.

Let G be the graph obtained from the HJ-construction by identifying v and v', removing an edge vz from K, removing an edge v'z' from G' and joining the vertices z and z'. Then G is of odd order, G is Δ-critical and contains two vertices u and u' of valency 2. Theorem 4.9 now follows from Lemma 4.8. //

Remarks.

1. It has been proved that there are no critical graphs of even order $p \leq 10$ and that there are no 3-critical graphs of order 12 and 14 (see Fiorini and Wilson [77]). Being prompted by these results, Jakobsen [74] made the following conjecture.

Critical Graph Conjecture : <u>There are no critical graphs of even order.</u>

(A similar conjecture was posed by Beineke and Wilson [73].)

2. M. K. Gol'dberg [79b] has constructed an infinite family of 3-critical graphs of even order, the smallest of which has order 22 which is shown in Fig.2.4. The subsequent graphs are obtained by adding pairs of 7-vertex blocks and joining the "hanging edges" to the base cycle as shown in Fig.2.10.

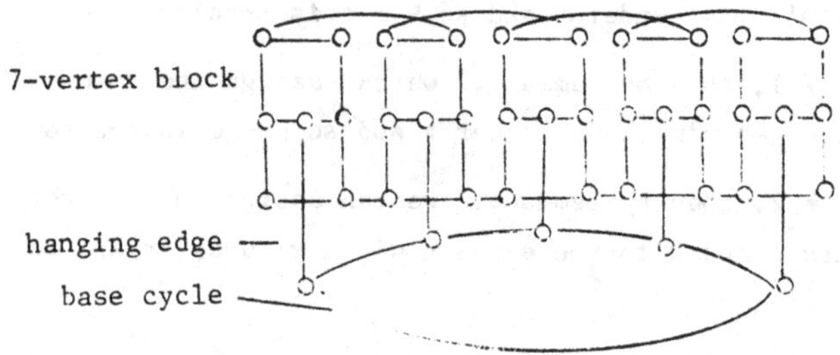

Figure 2.10

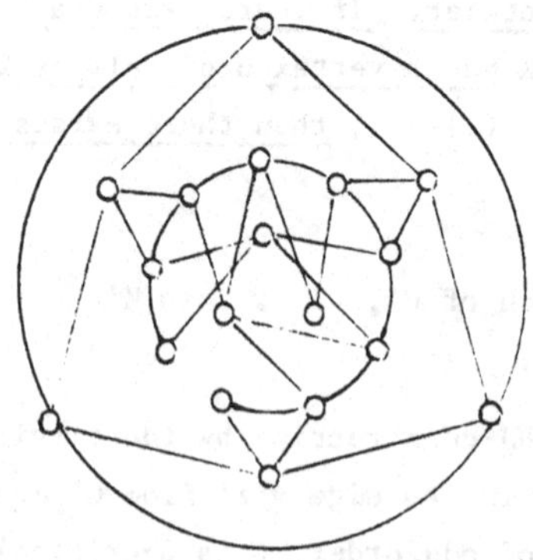

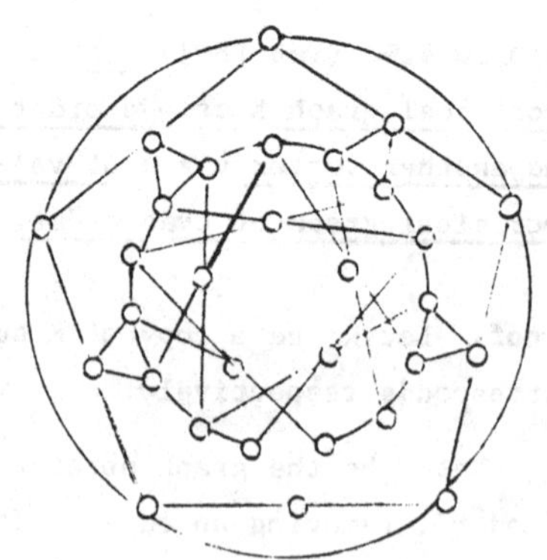

Figure 2.11

3. Fiol [80] independently found the counterexample to the Critical Graph Conjecture given in Fig.2.4. He also obtained two 4-critical graphs of order 18 and 30 which are shown in Fig.2.11 (quoted from Chetwynd and Wilson [83]).

4. A 4-critical graph of even order can be constructed by applying the graph given in Fig.2.5 to Theorem 4.9. However, Δ-critical graphs of even order for $\Delta > 5$ are still awaiting construction.

5. Gol'dberg [77] has proved that if G is a Δ-critical multigraph with $\chi'(G) > \frac{1}{8}(9\Delta + 6)$, then G is of odd order. This led him to formulate the following new conjecture.

Conjecture (Gol'dberg) There exists a constant $k > 2$ such that every Δ-critical multigraph with at most $k\Delta$ vertices is of odd order.

Exercise 2.4

1. Prove Lemma 4.7.
2. Prove that there are no critical graphs of order 4 and order 6.
3. Find all critical graphs of order 5.
4. Prove that there are no regular Δ-critical graphs for $\Delta > 3$.
5. For $i = 1,2,\ldots,t$, let G_i be a 3-critical graph obtained from C_{2s+1} by adding an independent set S_i of s edges. Prove that if $t < 2s - 4$, and if the sets S_i are pairwise disjoint, then the graph G obtained from C_{2s+1} by adding all of the sets S_i is a $(t+2)$-critical graph (Fiorini and Wilson [77;p.80]).
6. Let $\Delta > 3$ be an odd integer, let G be a Δ-critical graph and let $H = K_\Delta = \langle w_1,\ldots,w_\Delta \rangle$. Suppose v is a major vertex of G and $N(v) = \{v_1,\ldots,v_\Delta\}$. Let K be the graph obtained from G and H by deleting v from G and joining each v_i to w_i for all $i = 1,\ldots,\Delta$. Prove that K is Δ-critical and give an example showing that the result need not be true if Δ is even (Fiorini and Wilson [77]).
7. Let $\Delta > 3$ be an odd integer, let G be a Δ-critical graph and let $H = K_{\Delta,\Delta-1}$ having bipartition $\{w_1,\ldots w_{\Delta-1}\} \cup \{u_1,\ldots,u_\Delta\}$. Suppose v is a major vertex of G and $N(v) = \{v_1,\ldots,v_\Delta\}$. Let K be the graph

obtained from G and H by deleting v from G and joining each v_i to w_i for all $i = 1,\ldots,\Delta$. Prove that K is Δ-critical and give an example showing that the result need not be true if Δ is even (Fiorini and Wilson [77]).

8. Prove the following 'converse' of the result of Ex.3.4(6) : Let $\Delta > 3$ be an odd integer and let K be a Δ-critical graph which is separable by a set E of Δ independent edges into two graphs G and H. If each vertex in H is a major vertex in K, prove that the graph obtained from K by contracting H into a single vertex is also Δ-critical (Fiorini and Wilson [77]).

9. A graph G is vi-critical (vertex-index critical) if $\chi'(G - u) < \chi'(G)$ for each vertex u of G. A graph G is vc-critical (vertex-class critical) if G is of class 2, but G - u is of class 1 for each vertex u of G. Show that the graph G obtained from K_6 by inserting a new vertex into any edge is vi-critical but not vc-critical (Fiorini [78]).

10. Let H and K be the two graphs given in Fig.2.12. Let G be the graph obtained from H and K by identifying vertex u with vertex a, deleting the edges ux and ag from H and K respectively, and joining the vertices x and g by an edge. Let G^* be the graph obtained from G by joining the vertices v and y by an edge (see Fig.2.13). Show that G^* is not vi-critical (Fiorini [78]).

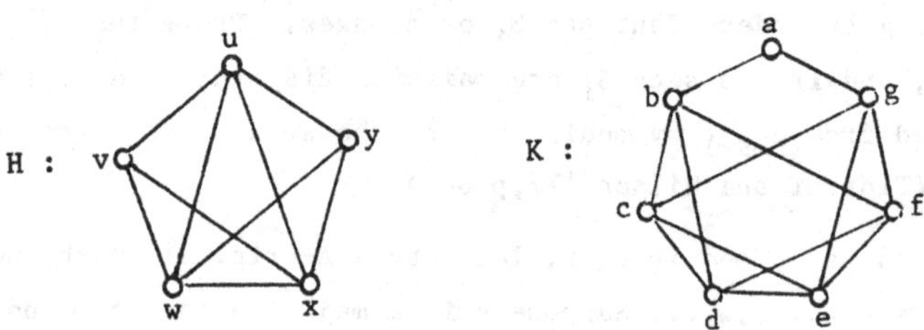

Figure 2.12

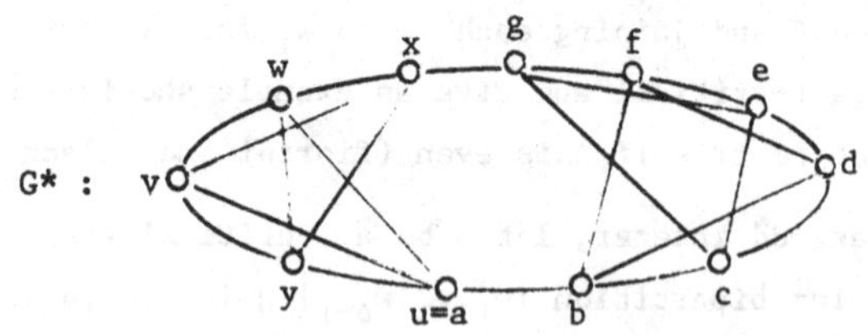

Figure 2.13

5. Bounds on the size of critical graphs

We shall now obtain some upper and lower bounds on the size of critical graphs. The main result of this section is an inequality involving the number of vertices n_i of valency i in a critical graph G due to Fiorini (Theorem 5.3). Using Fiorini's inequality, we are able to obtain very sharp lower bounds on the sizes of Δ-critical graphs for small Δ. We thus verify Vizing's conjecture on the lower bound for the size of Δ-critical graphs where Δ is small.

Theorem 5.1 below is taken from Fiorini and Wilson [77].

Theorem 5.1 _If_ G _is a_ Δ-_critical graph of order_ n _having minimum valency_ δ, _then_

$$e(G) \leq \begin{cases} \frac{1}{2}(n - 1)\Delta + 1 & \text{if } n \text{ is odd} \\ \frac{1}{2}(n - 2)\Delta + \delta - 1 & \text{if } n \text{ is even} \end{cases}$$

Proof. If n is odd, then $e(G) \leq \frac{1}{2}(n - 1)\Delta + 1$ follows immediately from Theorem 2.6.

If n is even, let v be a vertex of valency δ. By VAL, G has a major vertex v_1 adjacent to v. Let π be a Δ-colouring of $G - vv_1$. Then there is $i \in C(v) \setminus C(v_1)$. Let v_2 be such that $\pi(vv_2) = i$ and let G' be the multigraph obtained by adding an edge v_1v_2 in $G - v$. If v_1 and v_2 are joined by two edges coloured with colours i and j, we transform G' into a graph $\tilde{G}$ by replacing the double-edge v_1v_2 by a copy of $K_{\Delta,\Delta}$ with two non-adjacent edges removed as shown in Fig.2.14. Otherwise we let

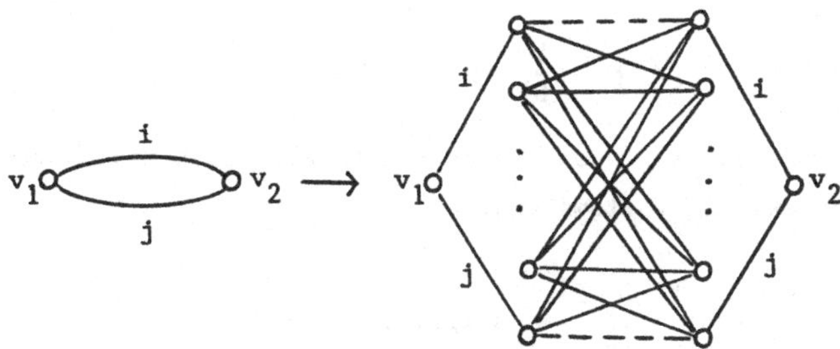

Figure 2.14

$\tilde{G} = G' + v_1v_2$. In either case, the graph $\tilde{G}$ is of class 1 and has odd order. It follows that, if $\tau(G)$ and $\tau(\tilde{G})$ are the total deficiencies of G and $\tilde{G}$ respectively, then $\tau(\tilde{G}) = \tau(G) - (\Delta - \delta) + (\delta - 2)$. However, by Corollary 2.7, we have $\tau(\tilde{G}) \geq \Delta$. Hence $\tau(G) \geq 2(\Delta - \delta + 1)$. Consequently, $e(G) \leq \frac{1}{2}(n - 2)\Delta + \delta - 1$. //

Corollary 5.2 *There are no regular* Δ-*critical graphs for* $\Delta \geq 3$.

Proof. By Corollary 2.7, there are no regular Δ-critical graphs of odd order if $\Delta \geq 3$. By Theorem 5.1, there are no regular critical graph of even order. //

Theorem 5.3 (Fiorini [75c]) *Let* $\Delta \geq 3$ *be an integer. If* G *is a* Δ-*critical graph, then*

$$n_\Delta \geq 2 \sum_{j=2}^{\Delta-1} \frac{n_j}{(j-1)}$$

where n_j *is the number of vertices of valency* j *in* G.

Proof. To each major vertex v in G, assign a $(\Delta-2)$-tuple $(i_2,\ldots,i_{\Delta-1})$, where i_t is the number of vertices of valency t adjacent to v. Let $n(i_2,\ldots,i_\Delta)$ be the number of major vertices of G associated with the $(\Delta-2)$-tuple $(i_2,\ldots,i_{\Delta-1})$. Let M be the set of all major vertices of G and for each $j = 2,\ldots,\Delta-1$, let A_j be the set of vertices of valency j in G. By counting the number of edges joining A_j and M in two different ways and noting that each vertex of G is adjacent to at least two major vertices, we have

$$2n_j \leq \sum_{(S)} i_j \, n(i_2,\ldots,i_{\Delta-1})$$

where the summation (S) extends over all $(\Delta-2)$-tuples associated with any major vertex. It follows that

$$\sum_{j=2}^{\Delta-1} \frac{2n_j}{j-1} \leq \sum_{j=2}^{\Delta-1} \sum_{(S)} \frac{i_j}{j-1} n(i_2,\ldots,i_{\Delta-1})$$

$$= \sum_{(S)} n(i_2,\ldots,i_{\Delta-1}) \sum_{j=2}^{\Delta-1} \frac{i_j}{j-1} \leq \sum_{(S)} n(i_2,\ldots,i_{\Delta-1}) \sum_{j=2}^{\Delta-1} \frac{i_j}{q-1}$$

where q is the smallest index of all non-zero elements of the $(\Delta-2)$-tuples $(i_2,\ldots,i_{\Delta-1})$. But by VAL, the vertex v is adjacent to at least $\Delta - q + 1$ major vertices and so it must be adjacent to at most $\Delta - (\Delta - q + 1) = q - 1$ minor vertices. Thus $i_2 + \ldots + i_{\Delta-1} \leq q - 1$, and so

$$2 \sum_{j=2}^{\Delta-1} \frac{n_j}{j-1} \leq \sum_{(S)} n(i_2,\ldots,i_{\Delta-1}) = n_\Delta. \qquad //$$

Corollary 5.4 *Suppose* G *is a* Δ-*critical graph of order* n *and size* m. *We have*

(i) *if* $\Delta = 3$, *then* $m \geq \frac{1}{4}(5n + 1)$;

(ii) *if* $\Delta = 4$, *then* $m \geq \frac{5}{3} n$; *and*

(iii) *if* $\Delta \geq 5$, *then* $m \geq n\left(\frac{3\Delta - 5}{\Delta}\right)$.

Proof. If $\Delta = 3$, then $2m = 2n_2 + 3n_3 = 2(n_2 + n_3) + \frac{1}{2} n_3 + \frac{1}{2} n_3 \geq 2n + \frac{1}{2} n_3 + n_2$. Hence $m \geq \frac{1}{4}(5n + 1)$.

If $\Delta = 4$, then $2m = 2n_2 + 3n_3 + 4n_4 \geq 2n_2 + 3n_3 + \frac{10}{3} n_4 + \frac{1}{3}(n_3 + 2n_2) = \frac{10}{3} n$. Hence $m \geq \frac{5}{3} n$.

For $\Delta \geq 5$,
$$\begin{aligned} 2m &= 2n_2 + 3n_3 + \ldots + \Delta n_\Delta \\ &\geq 4n_2 + 4n_3 + \ldots + (\Delta - 1)n_{\Delta-1} + (\Delta - 1)n_\Delta \\ &\geq 4(n - n_\Delta) + (\Delta - 1)n_\Delta = 4n + (\Delta - 5)n_\Delta. \end{aligned}$$

However, by VAL, each vertex of G is adjacent to at least two major vertices, and so $\Delta n_\Delta \geq 2n$. Part (iii) now follows by combining these two inequalities. //

The *valency-list* of a graph G is $1^{n_1} 2^{n_2} \ldots k^{n_k}$ where n_j is the number of vertices of valency j in G. If $n_j = 0$, the factor j^0 is usually omitted in the listing of the valency-list. The following theorem (Yap [81b]) answers a question raised by Jakobsen [74;p.269].

Theorem 5.5 *For each integer* $\Delta \geq 5$, *there are no* Δ-*critical graphs having valency-list* $2^r\Delta^{2r}$.

Proof. Suppose there is such a graph G. By VAL, each major vertex of G is adjacent to exactly one vertex of valency 2. Let x be a vertex of valency 2 and let $N(x) = \{y,z\}$, $e = xz$. Let π be a Δ-colouring of $H = G - e$ such that $\pi(xy) = \Delta$. By VAL, y and z are major vertices. Let

$$N(y) = \{x,y_1,\ldots,y_{\Delta-1}\}, \qquad N(z) = \{x,z_1,\ldots,z_{\Delta-1}\}.$$

If $z \in N(y)$, we let $z = y_{\Delta-1}$. We may assume that $\pi(yy_i) = i$, $\pi(zz_i) = i$, $i = 1,\ldots,\Delta-1$.

Now let w_i be the minor vertex adjacent to y_i (it may be possible that $w_j = w_i$ for some $j \neq i$) and let u_i be the other vertex adjacent to w_i. Certainly $z \neq u_i$ for any $i = 1,\ldots,\Delta-1$. Let $\pi(y_iw_i) = j_i$.

Now for each $i = 1,\ldots,\Delta-1$ let C_i be the $(\Delta,1)_\pi$-chain having origin x. By the Kempe-chain argument, the terminus of C_i must be z.

Suppose $j_1 = \Delta$. By the Kempe-chain argument on C_1, we will get a contradiction unless $\pi(w_1u_1) = 1$.

Suppose $j_1 \neq \Delta$. We can assume that $\pi(w_1u_1) \neq \Delta$. (If $\pi(w_1u_1) = \Delta$, we interchange the colours Δ and k ($\neq 1, j_1, \Delta$) in the $(\Delta,k)_\pi$-chain having origin w_1, and we get a new Δ-colouring λ of H such that $\lambda(w_1u_1) \neq \Delta$.) Then by the Kempe-chain argument on C_{j_1}, it is clear that

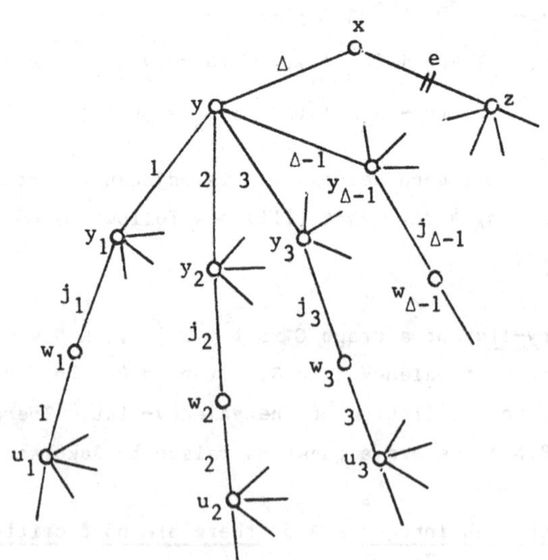

Figure 2.15

y_1, $w_1 \notin C_{j_1}$. We now interchange the colours in C_{j_1}. If $\pi(w_1u_1) = i \neq 1$, let Q be the $(j_1,1)$-chain having origin x in this new colouring of H. Then the terminus of Q must be w_1. By the Kempe-chain argument, this yields a contradiction. Hence $\pi(w_1u_1) = 1$ (see Fig. 2.15).

We have thus proved that $\pi(w_1u_1) = 1$ for any Δ-colouring π of H. Similarly, $\pi(w_iu_i) = i$, $i = 2,\ldots,\Delta - 1$ for any Δ-colouring π of H.

Finally, we consider the following cases.

Case 1. $j_1 \neq \Delta$ or $\Delta - 1$.

In this case, the terminus of the $(1,\Delta)_\pi$-chain C having origin w_1 cannot be y_1. Interchanging the colours in C we yield a contradiction to what we have proved above. Hence this case cannot occur.

Case 2. $j_1 = \Delta$ or $\Delta - 1$.

In this case, for each $i = 2,\ldots,\Delta$-2, let C_i be the $(1,i)_\pi$-chain having origin w_1. By the Kempe-chain argument, the terminus of C_i must be y_1. Hence y, $y_i \in C_i$. If $w_i \notin C_i$, then after interchanging the colours in C_i, yy_i and w_iu_i will receive distinct colours in this new colouring of H which has been shown impossible above. Hence $w_i \in C_i$ and $u_i \in C_i$ also. Thus $\pi(y_iw_i) = 1$ for each $i = 2,\ldots,\Delta$-2. Now consider the $(2,3)_\pi$-chain having origin w_2. By a similar argument, $\pi(y_3w_3) = 2$, which yields a contradiction. //

From the proof of Theorem 5.5, we also have

Theorem 5.6 _If_ G _is a_ 4-_critical graph such that_ $n_4 = 2n_2$ _and_ $n_3 = 0$, _then for each vertex_ x _of valency_ 2 _in_ G, _the two vertices adjacent to_ x _must be adjacent._

The result of Theorem 5.5 is not true for $\Delta = 3$. In fact, the graph given in Ex.2.3(4) (see Fig.2.3) is 3-critical and is such that $n_3 = 2n_2$.

Next, we apply Theorem 5.5, together with the proof of Theorem 5.3, to prove the following theorem (Yap [81b]).

Theorem 5.7 Let G be a Δ-critical graph of order n and size m.

(i) If $\Delta = 5$, then $m \geq 2n + 1$.

(ii If $\Delta = 6$, then $m \geq \frac{1}{4}(9n + 1)$.

(iii) If $\Delta = 7$, then $m \geq \frac{5}{2}n$.

Proof. Since the proofs of all the three parts are nearly identical, we prove only (i) here and leave (ii) and (iii) as exercises.

Let r be the number of vertices of valency 3 in G each of which is adjacent to a vertex of valency 4.

Similar to the proof of Theorem 5.3, we have

$$n_5 \geq 2n_2 + \frac{1}{2}\{3(n_3 - r) + 2r\} + \frac{2}{3}n_4, \quad r \leq \min\{n_3, n_4\}$$

$$= 2n_2 + n_3 + \frac{2}{3}n_4 + \frac{1}{2}(n_3 - r).$$

Hence

$$2m = 2n_2 + 3n_3 + 4n_4 + 5n_5 \geq 4(n_2 + n_3 + n_4 + n_5) + \frac{1}{2}(n_3 - r) + \frac{2}{3}n_4$$

which implies that

$$m \geq 2n + \frac{1}{4}(n_3 - r) + \frac{1}{3}n_4.$$

If either n_3 or n_4 is not 0, then $m \geq 2n + 1$. If $n_3 = n_4 = 0$, then by Theorem 5.5, $m \geq 2n + 1$. //

Vizing [68] made the following conjecture.

Vizing's conjecture on the lower bound for the size of critical graphs : Every Δ-critical graph of order n has size $m \geq \frac{1}{2}(n\Delta - n + 3)$.

Corollary 5.4, together with the results of Ex.2.4(2), 2.4(3), 2.6(3) and Theorem 6.9 verify Vizing's conjecture for $\Delta \leq 4$. The results of Theorem 5.7 are very close to the bounds given in Vizing's conjecture for $\Delta = 5$ and 6.

Exercise 2.5

1. Prove that if G is a Δ-critical graph, then

$$e(G) \geq \frac{1}{8}(3\Delta^2 + 6\Delta - 1) \qquad \text{(Vizing [65b]).}$$

2. Prove part (ii) and part (iii) of Theorem 5.7.

3. Assuming that Vizing's conjecture is true, prove that every planar graph G having $\Delta(G) \geq 7$ is necessarily of class 1 (Fiorini and Wilson [77;p.91]).

4. Prove that there are no 3-critical graphs of order 8 and 10.

5.* Let $1 \leq r \leq n$. Let G be a graph of order $2n + 1$ and $\Delta(G) = 2n + 1 - r$. Prove that G is of class 2 if and only if for some s such that $0 \leq s \leq (r-1)/2$ and for some set $\{v_1, v_2, \ldots, v_{2s}\} \subset V(G)$, $e(G - v_1 - \ldots - v_{2s}) > \binom{2n+1-2s}{2} - (r - 2s)(n - s)$ (Chetwynd and Hilton [84a]).

   (For $r = 1$, this conjecture was proved by Plantholt [81]. For $r = 2$, this conjecture was proved by Chetwynd and Hilton [84a].)

6. Suppose G is a graph of order 2n having r major vertices. Prove that if $\delta(G) \geq n + \frac{3}{2}r - 2$, then G is of class 1 (Chetwynd and Hilton [-b]).

6. Critical graphs of small order

In this section, we shall find all the critical graphs of order at most 8. We need the following definition. An *almost 1-factor* of a graph G of odd order n is a set consisting of $\frac{1}{2}(n - 3)$ independent edges of G.

Theorems 6.1 to 6.5 are due to Yap [81b]. We shall apply these general results later in this section.

Theorem 6.1 *Let G be a Δ-critical graph of order $n = \Delta + 1$ with minimum valency δ. If x is a vertex of valency δ in G and x is adjacent to r major vertices, then $\delta + r \geq \Delta + 2$.*

Proof. By VAL, we have $n_\Delta \geq \Delta - \delta + 2$. Now $n = \Delta + 1$ implies that $r = n_\Delta \geq \Delta - \delta + 2$. Hence $\delta + r \geq \Delta + 2$. //

Corollary 6.2 If G is a Δ-critical graph of order $n = \Delta + 1$ with minimum valency δ, then $\delta \geq \frac{1}{2}(\Delta + 2)$.

Theorem 6.3 For each odd integer $\Delta \geq 3$, there is exactly one Δ-critical graph of order $n = \Delta + 2$ with minimum valency 2, namely, the graph obtained from $K_{\Delta+1}$ by inserting a new vertex into an edge.

Proof. Suppose such a Δ-critical graph G exists. By VAL, $n_\Delta \geq n - 2$. By Ex.2.3(1), $n_\Delta \neq n - 2$. Hence $n_\Delta = n - 1$ and G is obtained from $K_{\Delta+1}$ by inserting a new vertex into an edge, which by Theroem 4.4, is Δ-critical. //

Theorem 6.4 For each even integer $\Delta \geq 4$, the graph G obtained from $K_{\Delta+1}$ by deleting an almost 1-factor is the only Δ-critical graph of order $n = \Delta + 1$ with minimum valency $\delta = \Delta - 1$.

Proof. Suppose G is a Δ-critical graph of odd order $n = \Delta + 1$ with minimum valency $\delta = \Delta - 1$. By Theorem 5.1, $e(G) \leq \frac{1}{2}(n - 1)\Delta + 1$. Let $\delta^{2r}\Delta^{n-2r}$ be the valency-list of G. Then

$$e(G) = \frac{1}{2}\{2r(\Delta - 1) + (n - 2r)\Delta\} = \frac{1}{2}(\Delta^2 + \Delta - 2r).$$

Hence $\Delta - 2r \leq 2$ and so $n_\Delta = n - 2r = (\Delta + 1) - 2r \leq 3$. However, by VAL, $n_\Delta \geq 3$. Combining these two inequalities, we have $n_\Delta = 3$ and $\Delta - 2r = 2$. Since the valency-list of G has been shown to be $\delta^{n-3}\Delta^3$, G is obtained from $K_{\Delta+1}$ by deleting an almost 1-factor.

Conversely, if G is obtained from $K_{\Delta+1}$ by deleting an almost 1-factor, then $e(G) > \Delta[\frac{n}{2}]$ and G is of class 2. If G is not critical, then G has a Δ-critical subgraph H of size less than $\frac{1}{2}\Delta^2 + 1$. By VAL, $3 \geq \Delta - \delta(H) + 2$. Hence $\delta(H) = \Delta - 1$. It is clear that $|H| = |G| = \Delta + 1$. But we have proved that the size of such a critical graph is $\frac{1}{2}\Delta^2 + 1$ which yields a contradiction. Hence G is critical. //

Theorem 6.5. For each odd integer $\Delta \geq 3$, the only possible valency-list of a Δ-critical graph G of order $n = \Delta + 2$ with minimum valency $\delta = \Delta - 1$ is $\delta^{n-4}\Delta^4$.

Proof. By VAL, $n_\Delta \geq 3$. Since Δ is odd and δ is even, $n_\Delta \geq 4$. Now by Theorem 5.1,

$$e(G) \leq \frac{1}{2}(n - 1)\Delta + 1 = \frac{1}{2}(\Delta + 1)\Delta + 1.$$

Suppose the valency-list of G is $\delta^{n-2r}\Delta^{2r}$, $r \geq 2$. Then

$$e(G) = \frac{1}{2}\{(n - 2r)(\Delta - 1) + 2r\Delta\} = \frac{1}{2}(\Delta^2 + \Delta + 2r - 2).$$

Hence $r = 2$ and the only possible valency-list of G is $\delta^{n-4}\Delta^4$. //

In the subsequent discussions, we need the following facts and definitions :

(1) From the fact that C_n, n odd, are the only 2-critical graphs and that there are no regular Δ-critical graphs for $\Delta \geq 3$ (Corollary 5.2), it follows that if G is a Δ-critical graph having minimum valency δ, then $2 \leq \delta < \Delta$.

(2) To show that a certain graph G is Δ-critical, sometimes we proceed as follows : we first show that G is of class 2 and contains no proper Δ-critical subgraph where $\Delta = \Delta(G)$. An argument of this kind will be called a *critical-list argument.*

(3) The graph obtained from K_4 by inserting a new vertex into an edge will be denoted by G_1.

We now apply the above results to prove Theorems 6.6 and 6.7.

Theorem 6.6 *C_5, G_1 and $K_5 - e$ where e is an edge of K_5, are the only three critical graphs of order* 5.

Proof. Suppose G is a Δ-critical graph of order 5 with minimum valency δ. If $\Delta = 2$, it is obvious that $G = C_5$. If $\Delta = 3$, by Theorem 6.3, $G = G_1$. If $\Delta = 4$, by Corollary 6.2, $\delta \geq 3$. Hence, by Corollary 5.2, $\delta = 3$. Finally, by Theorem 6.4, $G = K_5 - e$. //

Theorem 6.7 *There are no critical graphs of order* 4 *and* 6.

Proof. Suppose such a critical graph G exists. It is obvious that $\Delta = \Delta(G) \neq 2$. Hence we need only to consider the case that $\Delta \geq 3$.

Suppose $|G| = 4$. Then $3 \leq \chi'(G) \leq \chi'(K_4) = 3$. Hence G cannot be critical.

Suppose $|G| = 6$. If $\Delta(G) = 3$, then the only possible valency-list of G is $2^2 3^4$ which, by Ex.2.3.(1), is impossible. If $\Delta(G) = 4$, then applying VAL, we know that the possible valency-lists of G are 24^5 and $3^2 4^4$. However, by Theorem 5.1, the graphs having valency-lists 24^5 and $3^2 4^4$ cannot be critical. If $\Delta(G) = 5$, then $5 \leq \chi'(G) \leq \chi'(K_6) = 5$. Hence G cannot be critical. //

We shall apply the following lemma to prove Theorem 6.9.

Lemma 6.8 If G is a Δ-critical graph of odd order and F is a 1-factor of $G - x$ where x is a minor vertex of G, then $G - F$ has a $(\Delta-1)$-critical subgraph H.

Proof. $\chi'(G - F) = \Delta$, otherwise any $(\Delta-1)$-colouring of $G - F$ can be extended to a Δ-colouring of G. By the choice of x, $\Delta(G - F) = \Delta - 1$. Hence $G - F$ is of class 2. Lemma 6.8 now follows from Theorem 3.3. //

Theorem 6.9 (Beineke and Fiorini [76]) A 2-connected graph of order 7 is critical if and only if its valency-list is 2^7, 23^6, 24^6, $3^2 4^5$, 25^6, 345^5, $4^3 5^4$, $45^2 6^4$ or $5^4 6^3$.

Proof. Let G be a Δ-critical graph of order 7, size m and having minimum valency δ. By Theorem 3.2, G must be 2-connected. It should be clear by now that we need only to consider the cases that $\Delta \geq 3$ and $2 \leq \delta < \Delta = 6$.

$\Delta = 3$: By Fiorini's inequality, $n_3 \geq 2n_2$. Hence, the possible valency-list of G is 23^6. Now suppose G' is a graph having valency-list 23^6. Then by Theorem 2.6, G' is of class 2. If G' is not critical, then by Theorem 3.3, G' contains a 3-critical subgraph H. By Theorem 6.7, $|H| \neq 4,6$. Hence $|H| = 5$. However, if $|H| = 5$, then $H = G_1$. But G_1 cannot be extended to a graph having valency-list 23^6. (This argument is the so-called critical-list argument defined earlier.)

$\Delta = 4$: Suppose $\delta = 2$. By VAL and Fiorini's inequality, $n_4 \geq \max\{4, 2n_2 + n_3\}$. Hence the possible valency-lists of G are 24^6, $2^2 4^5$, and $23^2 4^4$. By Ex.2.3(1), $2^2 4^5$ is out. Applying VAL, we can see that the only possible 4-critical graph having valency-list 23^24^4 is the graph shown in Fig.2.16. However, this graph is 4-colourable. That

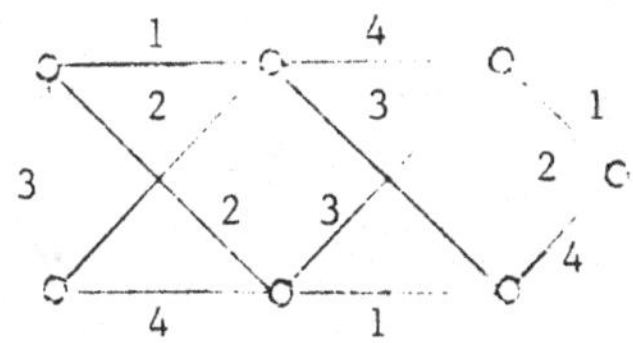

Figure 2.16

every graph having valency-list 24^6 is critical follows by a critical-list argument. Suppose $\delta = 3$. By Fiorini's inequality, $n_4 \geq n_3$. Hence 3^24^5 is the only possible valency-list of G. That every graph having valency-list 3^24^5 is critical follows by a critical-list argument.

Δ = 5 : Theorem 6.3 settles the case $\delta = 2$. Suppose $\delta = 3$. By now it should be easy to see that the possible valency-lists of G are 3^245^4 and 345^5. Suppose G is a graph having valency-list 3^245^4. Let x be a vertex of valency 3. Since $\chi'(G - x) = 5$ and $e(G - x) = 12$, $G - x$ has a 1-factor F. Thus, by Lemma 6.8, $G - F$ has a 4-critical subgraph H. It is clear that $|H| \neq 4,6,7$. Hence $H = K_5 - e$. Let $\{v_1,v_2\} = V(G) - V(H)$ and let B be the set of edges of G joining $\{v_1,v_2\}$ and V(H). It is clear that $e \notin E(G)$. Thus $F \subseteq B$, which is impossible. That a graph having valency-list 345^5 is critical follows from a critical-list argument. The case that $\delta = 4$ follows from Theorem 6.5 and a critical-list argument.

Δ = 6 : By Corollary 6.2, $\delta \geq 4$. Suppose $\delta = 4$. Then by VAL, $n_6 \geq 4$. Hence the possible valency-lists of G are 4^36^4, 45^26^4, 4^26^5 and 46^6. The valency-lists 46^6 and 4^26^5 are not realizable. If G is a graph having valency-list 4^36^4, then $G = K_7 - E(K_3)$ which is 6-colourable. That every graph having valency-list 45^26^4 is critical follows by a critical-list argument. Finally, for the case $\delta = 5$, we apply Theorem 6.4. //

Fig.2.17 gives all the Δ-critical graphs G of order 7 for $\Delta \geq 3$, except for the three graphs having valency-lists 25^6, 5^46^3 and 45^26^4. We may verify that all the graphs are non-isomorphic (for $\Delta \geq 4$) by considering their complements.

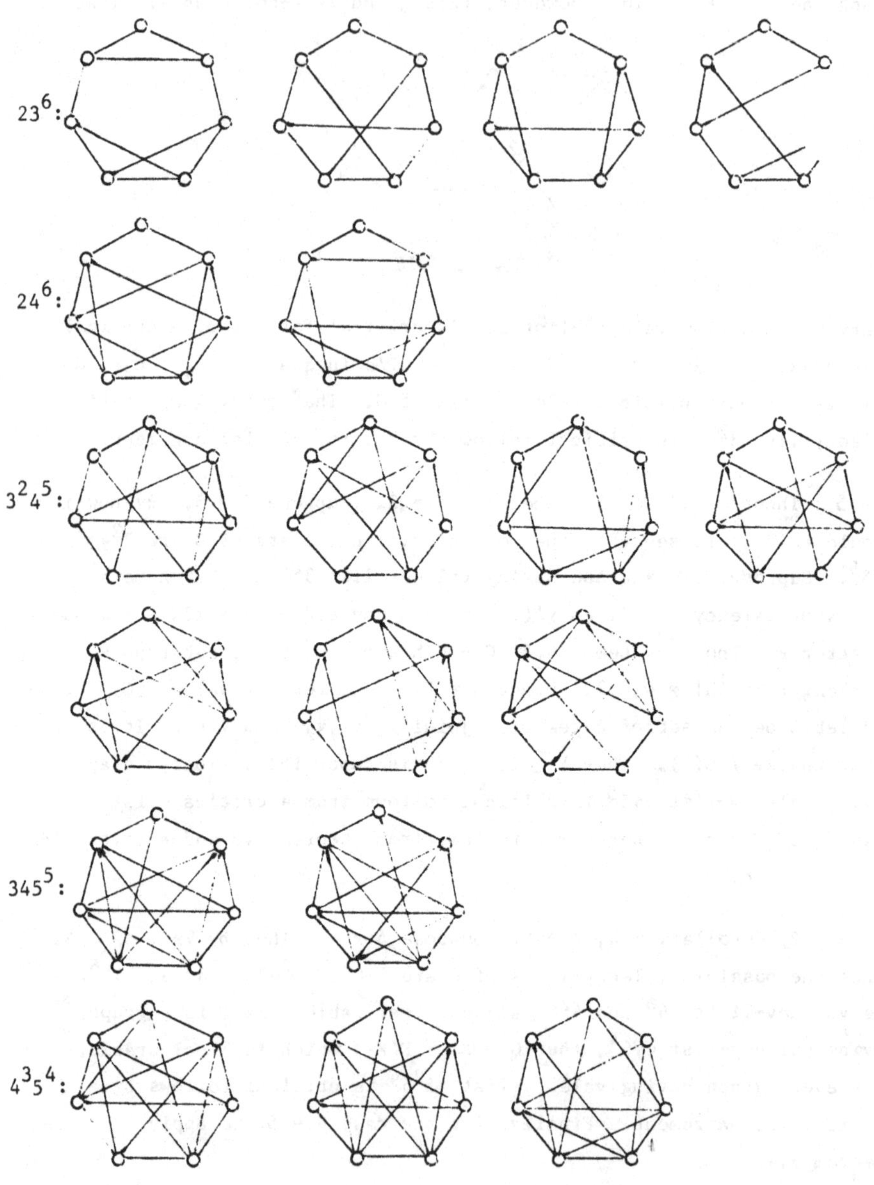

Figure 2.17

(Note that by Theorem 6.3, the graph obtained from K_6 by inserting a new vertex into an edge is the only critical graph having valency-list 25^6; by Theorem 6.4, the graph obtained from K_7 by deleting an almost 1-factor is the only graph having valency-list $5^4 6^3$; and it is clear that the graph obtained from K_7 by deleting two non-independent edges is the only graph having valency-list $45^2 6^4$.)

Exercise 2.6

1. Prove that if G is a Δ-critical graph of even order, then G has at least three minor vertices (Broere and Mynhardt [79]).
2. Applying Tutte's theorem on 1-factors, or otherwise, prove that if G is a critical graph of order 8 or 10, then G contains a 1-factor (Beineke and Fiorini [76]).
3. Prove that there are no critical graphs of order 8 (Beineke and Fiorini [76]).

4.+ Prove that there are no critical graphs of order 10 (Beineke and Fiorini [76]).

5. Prove that there are no 3-critical graphs of order 12 (Fiorini and Wilson [77;p.102]).

6.* Does there exist a critical graph of even order n where $12 < n < 16$?

7.+ Determine all 3-critical graphs of order 9 (Jakobsen [74]).

8.+ Prove that a 2-connected graph of order 9 having maximum valency $\Delta > 4$ is critical if and only if its valency-list is one of the following : 24^8, $3^2 4^7$ (except one graph), 25^8, 345^7, $4^3 5^6$, 26^8, 356^7, $4^2 6^7$, $45^2 6^6$, $5^4 6^5$, 27^8, 367^7, 457^7, $46^2 7^6$, $5^2 67^6$, $56^3 7^5$, $6^5 7^4$, $57^3 8^5$, $6^3 8^6$, $6^2 7^2 8^5$, $67^4 8^4$ and $7^6 8^3$ (Chetwynd and Yap [83]).

7. Planar graphs

In this section, we shall prove a result of Vizing (Theorem 7.1) and mention a long standing conjecture of Vizing [65b] concerning edge-colourings of planar graphs. The proof of Theorem 7.1 given below is due to Yap [81b]. It is a modification of a proof given by Mel'nikov [70].

Theorem 7.1 *Every planar graph whose maximum valency is at least* 8 *is necessarily of class* 1.

Proof. Suppose G is a planar graph of class 2 having $\Delta(G) > 8$. Then G contains an 8-critical subgraph. Hence, without loss of generality, we may assume that G is 8-critical.

From Euler's polyhedral formula, we have

$$e(G) < 3|G| - 6,$$

from which we can derive

$$12 + n_7 + 2n_8 < 4n_2 + 3n_3 + 2n_4 + n_5 \tag{1}$$

By VAL, a vertex of valency 3 is adjacent to at most one vertex of valency 7 (and is adjacent to no vertex of valency less than 7), a vertex of valency 4 is adjacent to at most one vertex of valency 6, and at most two vertices of valency 7. Let r be the number of vertices of valency 3 each of which is adjacent to a vertex of valency 7, let s be the number of vertices of valency 4 each of which is adjacent to a vertex of valency 6, let t be the number of vertices of valency 4 each of which is adjacent to exactly one vertex of valency 7, and let u be the number of vertices of valency 4 each of which is adjacent to exactly two vertices of valency 7.

Using the proof technique of Theorem 5.3, we have

$$n_8 > 2n_2 + \frac{1}{2}\{3(n_3 - r) + 2r\} + \frac{1}{3}\{4(n_4 - s - t - u) + 3s + 3t + 2u\}$$
$$+ \frac{1}{2}n_5 + \frac{2}{5}n_6 + \frac{1}{3}n_7$$

where $r < \min\{n_3, n_7\}$, $s < \min\{n_4, n_6\}$, $t < \min\{n_4 - s, n_7\}$ and $u < \min\{n_4 - s - t, n_7 - r\}$. Hence

$$n_7 + 2n_8 > 4n_2 + 3n_3 + 2n_4 + n_5 + (n_7 - r - u) + \frac{2}{3}(n_4 - s - t)$$
$$+ \frac{2}{3}(n_7 - u) > 4n_2 + 3n_3 + 2n_4 + n_5,$$

which contradicts (1). //

There exist planar graphs G with maximum valency $\Delta = 2,3,4,5$ which are either of class 1 or of class 2. Vizing [65a] made the following conjecture.

Planar Graph Conjecture. *If* G *is a planar graph whose maximum valency is at least* 6, *then* G *is of class* 1.

The following theorems (Yap [81b]) show that if a 7-critical or 6-critical planar graph G exists, then G has quite a few minor vertices. These results may be of some use in settling the Planar Graph Conjecture.

Theorem 7.2 *If a* 7-*critical planar graph* G *exists, then*

$$2n_3 + \frac{4}{3}n_4 + \frac{1}{2}n_5 > 12 + \frac{2}{5}n_6 \quad \underline{\text{and}} \quad n_7 > 6 + 2n_2 + \frac{1}{4}n_5 + \frac{3}{5}n_6.$$

Proof. Let S be the set of all vertices of valency 2 in G. Let m_j be the number of vertices of valency j in G - S. From Euler's polyhedral formula, we can derive

$$4m_2 + 3m_3 + 2m_4 + m_5 > 12 + m_7 \tag{2}$$

By VAL, $m_2 = 0$, $m_j = n_j$, $j = 3,4,5$ and $m_7 + 2n_2 = n_7$. Hence (2) becomes

$$2n_2 + 3n_3 + 2n_4 + n_5 > 12 + n_7 \ .$$

Now applying Fiorini's inequality $n_7 > 2n_2 + n_3 + \frac{2}{3}n_4 + \frac{1}{2}n_5 + \frac{2}{5}n_6$, we have $2n_3 + \frac{4}{3}n_4 + \frac{1}{2}n_5 > 12 + \frac{2}{5}n_6$ and $n_7 > 6 + 2n_2 + \frac{1}{4}n_5 + \frac{3}{5}n_6$. //

Corollary 7.3 *If a* 7-*critical planar graph* G *exists, then*

$$n_3 + n_4 + n_5 > 6 \quad \underline{\text{and}} \quad n_7 > 6 + 2n_2.$$

Theorem 7.4 *If a* 6-*critical planar graph* G *exists, then*

$$2n_2 + 3n_3 + 2n_4 + n_5 > 12 \quad \underline{\text{and}} \quad n_6 > 4 + \frac{4}{3}n_2 + \frac{1}{6}n_5.$$

The proof of this theorem is similar to that of Theorem 7.2 and thus is left as an exercise.

Exercise 2.7

1. Prove Theorem 7.4.

2. A planar graph G is <u>outerplanar</u> if it can be embedded in the plane in such a way that G has no crossings and that all its vertices lie on the boundary of the same face. Prove that an outerplanar graph is of class 1 if and only if it is not a cycle of odd length. (This result is due to C. McDiarmid, see Fiorini and Wilson [77;p.109].)

8. 1-factorization of regular graphs of high degree

A regular graph of even order is said to be <u>1-factorizable</u> if its edge-set is the union of 1-factors. The following well-known conjecture probably was made independently by many people.

1-Factorization Conjecture <u>If a regular graph</u> G <u>of order</u> 2n <u>has degree</u> deg $G \geq 2[\frac{n+1}{2}] - 1$, <u>then</u> G <u>is</u> 1-<u>factorizable</u>.

It is clear that if G is of even order, then G is 1-factorizable if and only if G is of class 1. Thus Theorem 1.2 verifies that this conjecture is true for deg G = 2n - 1. If deg G = 2n - 2, then G is obtained from K_{2n} by deleting a 1-factor. However, it is clear that any 1-factor in K_{2n} can be chosen as a colour class in a (2n-1)-colouring of K_{2n}. Thus this conjecture is also true for deg G = 2n - 2.

The result of Ex.2.8(1) shows that if this conjecture is true then the lower bound $2[\frac{n+1}{2}] - 1$ is best possible.

Rosa and Wallis [82] proved this conjecture for the case deg G = 2n - 4 under the assumption that $\bar{G}$ is also 1-factorizable. Chetwynd and Hilton [85] proved this conjecture for deg G ≥ 2n - 5. As their proof for the case that deg G = 2n - 5 is complicated, we shall only reproduce their proof here for the case that deg G ≥ 2n - 4 and leave the proof for the case that deg G = 2n - 5 as an exercise.

We shall require the following lemmas.

Lemma 8.1 <u>Suppose</u> G <u>is a regular graph of order</u> 2n <u>and</u> $G \neq K_{2n}$. <u>Then</u> G <u>is of class</u> 1 <u>if and only if for any</u> $w \in V(G)$, G - w <u>is of class</u> 1.

Proof. Necessity. Since G is regular and $G \neq K_{2n}$, $\Delta(G - w) = \Delta(G)$ for any $w \in V(G)$. Hence $\Delta(G - w) < \chi'(G - w) < \chi'(G) = \Delta(G) = \Delta(G - w)$, from which it follows that $\chi'(G - w) = \Delta(G - w)$ and $G - w$ is of class 1.

Sufficiency. Let $\Delta = \Delta(G - w)$ and let π be a Δ-colouring of $G - w$. If there is a colour, colour i say, which is absent at more than one vertex in $N(w)$, then there is a colour, colour j say, which is present at all the vertices in $N(w)$ and thus colour j is present at every vertex in $G - w$. But this is impossible because $G - w$ has an odd number of vertices. Thus each colour is absent from exactly one vertex in $N(w)$ and π can obviously be extended to a Δ-colouring of G. //

Lemma 8.2 For a graph G, let $e = vw \in E(G)$. If w is adjacent to at most one vertex of valency $\Delta(G)$, then

$$\Delta(G - e) = \Delta(G) \Rightarrow \chi'(G - e) = \chi'(G), \quad \text{and}$$

$$\Delta(G - w) = \Delta(G) \Rightarrow \chi'(G - w) = \chi'(G).$$

Proof. If G is of class 1, then $\Delta(G) = \chi'(G) > \chi'(G - e) > \Delta(G - e) = \Delta(G)$ and thus $\chi'(G) = \chi'(G - e)$. Similarly, $\chi'(G - w) = \chi'(G)$.

If G is of class 2, then by Theorem 3.3, G contains a $\Delta(G)$-critical subgraph G^*. Now by VAL, $w \notin V(G^*)$. Thus $\chi'(G - w) = \chi'(G)$ and $\chi'(G - e) = \chi'(G)$. //

Lemma 8.3 Let G be a graph of order n and let $\Delta = \Delta(G) > 3$. Suppose G has three vertices of valency Δ. Then G is of class 2 if and only if $\Delta = n - 1$ and each of the remaining vertices has valency $n - 2$ (and thus n is odd).

Proof. Sufficiency. Suppose G has three vertices of valency $n - 1$ and each of the remaining vertices has valency $n - 2$. Then $2e(G) = 3(n - 1) + (n - 3)(n - 2) = n(n - 2) + 3$, from which it follows that n is odd. Hence $e(G) > \Delta[\frac{n}{2}]$ and by Theorem 2.6, G is of class 2.

Necessity. Suppose G has three major vertices (a,b and c say) and is of class 2. Since $\Delta > 3$, $n > 5$. By Theorem 3.3, G contains a Δ-critical subgraph G^*. Since there are no regular Δ-critical graphs for $\Delta > 3$ (see Ex.2.4(4)), by VAL, G^* has the same three major vertices a, b and c, and $\delta(G^*) > \Delta - 1$. Thus $\delta(G^*) = \Delta - 1$, $G^* = G$ and n is odd. Now for $n \leq 9$, the necessity is true (this follows from Theorems 6.6, 6.9 and Ex.2.6(7) and 2.6(8)). Hence we assume $n \geq 11$. By Theorem 5.3, $\Delta > \frac{2}{3}n$ and thus for $n \geq 11$, $\delta(G - a - b) \geq \Delta - 3$ $> \frac{2}{3}n - 3 \geq \frac{1}{2}(n - 2)$. Consequently, by Dirac's theorem, $G - a - b$ has a Hamilton circuit. Now if $\Delta < n - 1$, then G has a vertex $d \notin N(a)$ such that G has a near 1-factor F (a <u>near 1-factor</u> of a graph G of odd order n is a set of $\frac{1}{2}(n - 1)$ independent edges of G) which contains the edge ab but does not include any edge incident with d. Thus $G - F$ has four vertices a, b, c, d of maximum valency $\Delta - 1$, joined as illustrated in Fig.2.18.

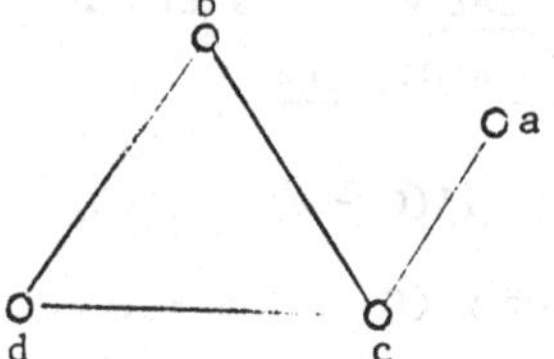

Figure 2.18

Since a is adjacent to only one vertex of maximum valency $\Delta - 1$ in $G - F$ and $\Delta(G - F - ac) = \Delta(G - F)$, by Lemma 8.2, $G - F$ and $G - F - ac$ are of the same class. However, since $G - F - ac$ has only two vertices of maximum valency, $G - F - ac$ is of class 1. Hence G is of class 1. This contradiction proves the necessity. //

Theorem 8.4 <u>Suppose</u> G <u>is a regular graph of order</u> 2n <u>and</u> deg $G = 2n - 3$ <u>or</u> $2n - 4$. <u>Let</u> deg $G \geq 2[\frac{n+1}{2}] - 1$. <u>Then</u> G <u>is</u> <u>1-factorizable</u>.

Proof. Suppose deg $G = 2n - 3$. Let $w \in V(G)$. Then the graph $G - w$ has only two major vertices and thus by VAL, $G - w$ is of class 1. Hence, by Lemma 8.1, G is 1-factorizable.

Suppose deg $G = 2n - 4$. Clearly deg $G = 2n - 4 \geq 2[\frac{n+1}{2}] - 1$ implies that $n \geq 4$. Thus for any $w \in V(G)$, $G - w$ has three vertices of valency $\Delta(G) \geq 4$. Now by Lemma 8.3, $G - w$ is of class 1 and thus by Lemma 8.1, G is 1-factorizable. //

Exercise 2.8

1. Let G be the graph obtained from two copies of K_{2m+1}, where $m > 2$, by deleting one edge (say a_1b_1 and a_2b_2) from each, and joining them by the two edges a_1a_2 and b_1b_2. Prove that G is of class 2 (Chetwynd and Hilton [85]).

2.$^+$ Suppose G is a regular graph of order 2n and deg G = $2n - 5 > 2[\frac{n+1}{2}] - 1$. Prove that G is 1-factorizable (Chetwynd and Hilton [85] and [-c]).

3.$^+$ Prove that the 1-Factorization Conjecture is true for deg G $> \frac{6}{7}|V(G)|$ (Chetwynd and Hilton [85]).

4. We define the graph G_n as follows. The vertex set of G_n is the symmetric group Σ_n on $\{1,2,\ldots,n\}$. Two vertices δ, τ in G_n are adjacent if and only if $\delta^{-1}\tau$ has exactly one nontrivial cycle. Then G_n is a Cayley graph (see Chapter 3). Prove that G_n is 1-factorizable (Brualdi [78]).

5.* Let $1 \leq r \leq n$ and let G be a graph of order 2n + 2 with $\Delta(G) = 2n + 1 - r$. Prove that G is of class 2 if and only if for some s such that $0 \leq s \leq (r-1)/2$ and for some set $\{v_1, v_2, \ldots, v_{2s+1}\} \subset V(G)$, $e(G - v_1 - \ldots - v_{2s+1}) > \binom{2n+1-2s}{2} - (r - 2s)(n - s)$ (Chetwynd and Hilton [84b]).

(For r = 1, this conjecture was proved by Plantholt [83]. For r = 2, this conjecture was proved by Chetwynd and Hilton [84b].)

6.* **The Odd Graph Conjecture** For each integer $k \geq 2$, the odd graph O_k^* is the graph obtained by taking as vertices each of the (k-1)-subsets of $\{1, 2, \ldots, 2k - 1\}$, and joining two of these vertices with an edge whenever the corresponding subsets are disjoint. Thus all odd graphs are regular. Prove that O_k^* is of class 1 unless k = 3 or $k = 2^r$ for some integer r (Fiorini and Wilson [77;p.45]).

7.$^+$ Suppose G is a graph having four major vertices. Prove that G is of class 2 if and only if for some integer n, one of the following holds :

(i) the valency-list of G is $(2n - 2)^{2n-3}(2n - 1)^4$;

(ii) the valency-list of G is $(2n - 2)(2n - 1)^{2n-4}(2n)^4$;

(iii) G contains a bridge e such that $G - e$ is the union of two graphs G_1 and G_2 where $\Delta(G_1) \leq 2m - 1$ for some integer $m < n$ and the valency-list of G_2 is either $(2m - 2)(2m - 1)^{2m-4}(2m)^4$ or $(2m - 1)^{2m-2}(2m)^3$ (Chetwynd and Hilton [85]).

8. Suppose G is a regular graph of even order. Prove that if G is of class 1, then the line graph of G is also of class 1 (Jaeger [73]).

9. Applications to vertex-colourings

Several applications of edge-colourings to electrical networks, scheduling problems, constructions of latin squares etc., can be found in many existing books on graph theory. In this section, we shall present some recent results on nontrivial applications of the theory of edge-colourings to find a very sharp upper bound for the chromatic number $\chi(G)$ of graphs G that do not induce $K_{1,3}$ or $K_5 - e$ (the complete graph K_5 minus an edge). These results are due to S. A. Choudum, M. Javdekar, H. A. Kierstead and J. Schmerl.

It can be easily shown that for any graph G, $\chi(G) \leq \Delta(G) + 1$. R. L. Brooks (1941) strengthened this result by showing that for any connected graph G which is neither an odd cycle nor a complete graph, $\chi(G) \leq \Delta(G)$. The upper bound for $\chi(G)$ obtained by Brooks' is not sharp, for instance, $\chi(K_{1,n}) = 2$ whereas $\Delta(K_{1,n}) = n$; for any planar graph G, $\chi(G) \leq 4$ while $\Delta(G)$ can be arbitrarily large. Other upper bounds for $\chi(G)$ are also known, but they are also not sharp enough.

Given a graph G, we denote its <u>density</u> (<u>clique number</u>), the maximum number of vertices in a clique, by $\omega(G)$. Suppose π is a proper vertex-colouring of G. Let $G_{i,j}$ be the subgraph of G induced by the vertices coloured with colours i and j. A vertex u such that $\pi(u) = k$ is called a <u>k-vertex</u>. We shall follow the convention that if $H = \langle u_1, u_2, u_3, u_4 \rangle \simeq K_{1,3}$, then $d_H(u_1) = 3$ and $d_H(u_i) = 1$ for $i \geq 2$.

Let L(G) be the line graph of a graph G. It is easy to see that $\chi'(G) = \chi(L(G))$ and if $G \neq K_3$, $\Delta(G) = \omega(L(G))$. In view of these facts and a theorem of Beineke [68] characterizing line graphs in terms of nine forbidden subgraphs (see Fig.2.19), the following result follows as a consequence of Vizing's theorem : If a graph G does not induce any of the nine graphs in Fig.2.19, then $\omega(G) \le \chi(G) \le \omega(G) + 1$.

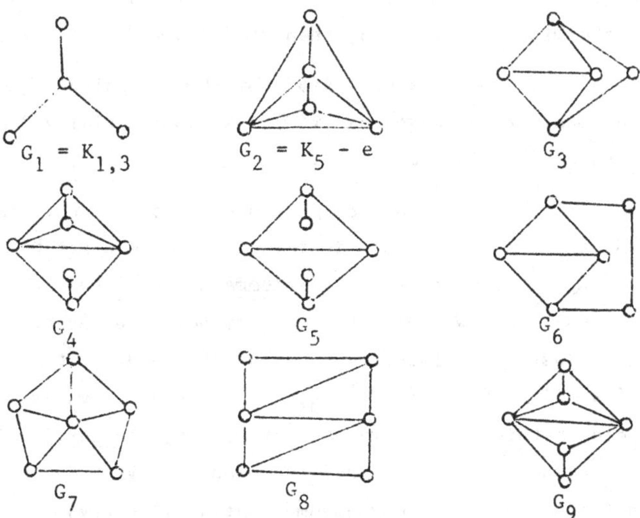

Figure 2.19

Choudum [76] showed that the conclusion of the above result is true for a wider class of graphs, namely, if G does not induce G_1, G_2, G_3 and G_4 or if G does not induce G_1, G_2, G_5 and G_6, then $\omega(G) \le \chi(G) \le \omega(G) + 1$. Javdekar [80] improved Choudum's results by eliminating G_3 and G_5 from the hypothesis. He further conjectured that if a graph G does not induce G_1 and G_2, then $\omega(G) \le \chi(G) \le \omega(G) + 1$. Kierstead and Schmerl [83] proved that Javdekar's conjecture is true for graphs G with $\omega(G) \le 9$. They also showed that if G does not induce any of the graphs G_1, G_2 or all of G_4, G_5 and G_9, then $\omega(G) \le \chi(G) \le \omega(G) + 1$. Kierstead [84] has now proved Javdekar's conjecture.

To prove Javdekar's conjecture, we need the following lemma (due to Choudum [77]) :

Lemma 9.1 *If a graph* G *does not induce* $K_{1,3}$ *and* $\Delta(G) \le 5$, *then*

$$\chi(G) \le \omega(G) + 1.$$

Proof. We need only to consider the case that G is connected. Let $\Delta = \Delta(G)$. Clearly this lemma is true if $\Delta \le 2$. We first note that for

$\Delta \geq 3$, since G is claw-free, i.e. G does not induce $K_{1,3}$, $\omega(G) \geq 3$.

Suppose $\Delta = 3$. Then $3 \leq \omega(G) \leq 4$. If $\omega(G) = 4$, then $G = K_4$ and $\chi(G) = 4$. Otherwise, if $\omega(G) = 3$, then $\chi(G) \leq \Delta + 1 = \omega(G) + 1$.

Suppose $\Delta = 4$. Then $3 \leq \omega(G) \leq 5$. By the foregoing argument, we need only to consider the case that $\omega(G) \leq 4$. Now applying Brooks' theorem, we have $\chi(G) \leq \Delta \leq \omega(G) + 1$.

Suppose $\Delta = 5$. By the foregoing argument, we need only to consider the case $\omega(G) = 3$. If the result is false, let G be a graph of minimum order satisfying the hypothesis of the lemma and such that $\Delta(G) = 5$, $\omega(G) = 3$ and $\chi(G) = 5$. We note that for any vertex u having valency 5, $\langle N(u)\rangle \cong C_5$ because G is claw-free $K_{1,3}$ and $\omega(G) = 3$. Let v be a vertex of valency 5 and let $N(v) = \{v_1,v_2,v_3,v_4,v_5\}$ where v_1, v_2, v_3, v_4 and v_5 form a 5-cycle in this order. By the assumption, for any 4-colouring π of $G - v$, the number of colours used to colour the vertices $v_1, \ldots, v_5$ must be exactly 4. Hence we may assume that $\pi(v_1) = \pi(v_3) = 1$, $\pi(v_2) = 2$, $\pi(v_4) = 3$ and $\pi(v_5) = 4$. Now in $G - v$, there exist two paths $P(v_2,v_4)$ and $P(v_2,v_5)$ in $G_{2,3}$ and $G_{2,4}$ respectively; otherwise by interchanging the colours of a component, as in the proof of the well-known 5-Colour Theorem, we get a 3-colouring of $\langle N(v)\rangle$, a contradiction. Let $v_6 \in P(v_2,v_4)$ and $v_7 \in P(v_2,v_5)$ be the vertices coloured with colour 2 and adjacent to v_4 and v_5 respectively. (We shall later consider the cases that $v_7 \neq v_6$ and $v_7 = v_6$ separately.) Let v_8 and v_9 be the other two vertices adjacent to v_2. Then $\{\pi(v_8),\pi(v_9)\} = \{3,4\}$. We assume that the 5-cycle induced by $N(v_2)$ is in the order v_9, v_3, v, v_1.

Case 1. $v_7 \neq v_6$.

First suppose $\pi(v_8) = 4$ and $\pi(v_9) = 3$ (see Fig.2.20). Now either

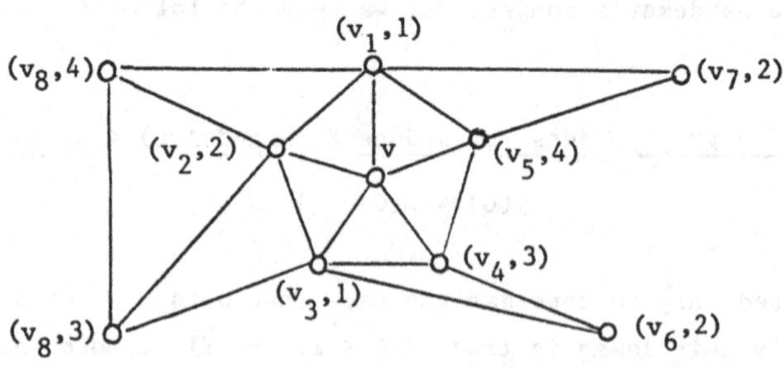

Figure 2.20

$v_6v_3 \in G$ or $v_6v_5 \in G$ (otherwise $\langle v_4,v_3,v_5,v_6\rangle \simeq K_{1,3}$). If $v_6v_5 \in G$, then $\langle v_5,v,v_6,v_7\rangle \simeq K_{1,3}$. Hence $v_6v_3 \in G$. Similarly $v_7v_1 \in G$. Now $d(v_1) = 5 = d(v_3)$ and there is no 4-vertex in the neighbourhood of v_3. Hence, after recolouring v_3 with colour 4, we get a 3-colouring of $N(v)$, a contradiction.

Next suppose $\pi(v_8) = 3$ and $\pi(v_9) = 4$ (see Fig.2.21). As above,

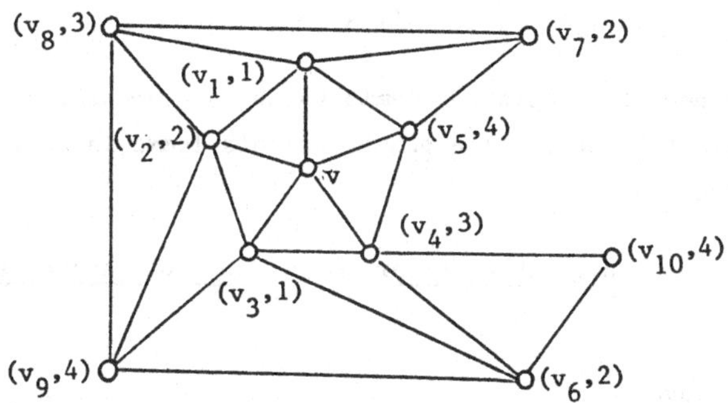

Figure 2.21

$v_6v_3 \in G$ and $v_7v_1 \in G$. Consequently v_6v_9, $v_7v_8 \in G$. But then $\langle v_2,v_9,v_6\rangle$ is a subpath of $P(v_2,v_5)$ in $G_{2,4}$ because for any colour k, there are at most two k-vertices adjacent to a vertex (otherwise G has an induced $K_{1,3}$). Hence there is a 4-vertex, v_{10} say, adjacent to v_6. Since $\omega(G) = 3$, $v_4v_9 \notin G$ and so $v_{10}v_4 \in G$ (otherwise $\langle v_6,v_9,v_4,v_{10}\rangle \simeq K_{1,3}$). Now $\langle v_4,v_3,v_5,v_{10}\rangle \simeq K_{1,3}$, a contradiction.

Case 2. $v_7 = v_6$.

This case can be settled in a similar way. We shall leave it as an exercise. //

Let M be a multigraph. A triangle with a multiple edge in M is called a 4-sided triangle in M. Let G be a graph and $v \in V(G)$. The closed neighbourhood $N[v]$ of v is $N(v) \cup \{v\}$.

Kierstead and Schmerl [83] proved that the following two assertions are equivalent.

G(n) : If G is a graph that does not induce $K_{1,3}$ or $K_5 - e$ and $\omega(G) \leq n$, then $\chi(G) \leq n + 1$.

$M(n)$: If M is a multigraph such that $\Delta(M) \le n$, $\mu(M) \le 2$ and M does not contain a 4-sided triangle, then $\chi'(M) \le n + 1$.

To prove that $G(n)$ and $M(n)$ are equivalent, we shall require part of Corollary 3.5 which we now restate it as Lemma 9.2 below.

Lemma 9.2 *If* M *is a chromatic-index-critical multigraph such that* $q = \chi'(M) \ge \Delta(M) + 1$, *then every vertex* v *of* M *is adjacent to at least two vertices* x *such that* $d(x) + \mu(v,x) = q$.

We also need the following lemma which is a special case of a well-known theorem of Ramsey. (The proof for this lemma is easy and is left as an exercise.)

Lemma 9.3 *If* G *is a graph of order* $n \ge 6$, *then* G *contains either* K_3 *or* O_3.

We now prove

Theorem 9.4 (Kierstead and Schmerl [83]) *For every positive integer* n, $G(n)$ *holds if and only if* $M(n)$ *holds*.

Proof. We shall apply the following three lemmas to prove this theorem. Lemma 9.5 states that both $G(n)$ and $M(n)$ hold for $n \le 3$.

We now prove that for $n \ge 3$, $G(n)$ implies $M(n)$. Let M be a multigraph satisfying the hypothesis of $M(n)$ and let G be the line graph of M. By Lemma 9.6, G satisfies the hypothesis of $G(n)$ and thus $\chi(G) \le n + 1$. Consequently $\chi'(M) = \chi(G) \le n + 1$ and $M(n)$ holds.

Finally, Lemma 9.7 shows that for $n \ge 4$, if a graph G is a counter-example to $G(n)$, then there exists a multigraph M which is a counter-example to $M(n)$. Hence $M(n)$ implies $G(n)$. //

Lemma 9.5 *If* $n \le 3$, *then both* $G(n)$ *and* $M(n)$ *are true.*

Proof. Suppose M is a multigraph with $\Delta(M) \le n \le 3$. Then by Shannon's theorem, $\chi'(M) \le [\frac{3}{2}\Delta] \le [\frac{3}{2}n] \le n + 1$. Hence, for $n \le 3$, $M(n)$ is true (even if M contains a 4-sided triangle).

Suppose G is a graph that does not induce $K_{1,3}$ and $\omega(G) \le n \le 3$. It is clear that for $n \le 2$, $\chi(G) \le 3$. Suppose $n = 3$. If $\Delta(G) \ge 6$, let

$v \in G$ be such that $d(v) \geq 6$ and let $H = \langle N(v) \rangle$. By Lemma 9.3, either H contains K_3 and thus $\omega(G) \geq 4$ or H contains O_3 and thus G induces $K_{1,3}$. Hence $\Delta(G) \leq 5$ and this lemma follows from Lemma 9.1. //

In general if M is a multigraph and G is the line graph of M, then G does not induce $K_{1,3}$ and $\omega(G) = \Delta(M)$. Furthermore, if $\mu(M) \leq 2$, then G does not induce $K_5 - e$. Hence we have the following lemma in which only the condition that $\mu(M) \leq 2$ is required and for which G does not induce $K_{1,3}$ is always true.

Lemma 9.6 *If M is a multigraph such that $\Delta(M) \geq 3$, $\mu(M) \leq 2$, and M has no 4-sided triangles, then the line graph G of M does not induce $K_{1,3}$ or $K_5 - e$ and $\omega(G) = \Delta(M)$.*

Next we prove

Lemma 9.7 *If $G = (V,E)$ is a chromatic-critical graph such that G does not induce $K_{1,3}$ or $K_5 - e$, $\omega(G) \geq 4$, and $\chi(G) \geq \omega(G) + 2$, then G is the line graph of a multigraph M such that $\Delta(M) = \omega(G)$, $\mu(M) \leq 2$, and M does not contain a 4-sided triangle.*

Proof. Let $M = (W,F)$ where

$W = \{X \subseteq V \mid X$ is a clique and for some clique Y in G, $X \cap Y \neq \phi$,

$\min\{|X - Y|, |Y - X|\} \geq 2$ and $\max\{|X|, |Y|\} \geq 4\}$ and

$F = \{XY_v \mid X, Y \in W, X \neq Y$, and $v \in X \cap Y\}$

(Here XY_v denotes the edge joining X and Y which is labelled by v.)

Let $\psi : F \to V$ be defined by $\psi(XY_v) = v$. We shall now prove that G is the line graph of M by showing that ψ is bijective and that two edges in M are adjacent if and only if their images are adjacent in G. First we establish some useful assertions about G.

(A1) *If K is a clique and $v \notin K$, then $|N(v) \cap K| \leq 2$.*

Suppose $|N(v) \cap K| \geq 3$. Since $v \notin K$, there exists $k \in K$ such that $vk \notin E$. Let $A \subseteq N(v) \cap K$ be such that $|A| = 3$, then $\langle A \cup \{v,k\} \rangle \cong K_5 - e$, a contradiction. Hence $|N(v) \cap K| \leq 2$.

(A2) *If K and K' are distinct cliques in G, then $|K \cap K'| \leq 2$.*

Since $K' \neq K$, there exist $v \in K'$ and $k \in K$ where $vk \notin E$. If

$|K \cap K'| \geq 3$, let $A \subseteq K \cap K'$ be such that $|A| = 3$. Then $\langle A \cup \{v,k\} \rangle \simeq K_5 - e$, a contradiction.

(A3) *If* K *is a clique in* G, $|K| \geq 4$, *and* $k \in K$, *then* $\langle (N(k) - K) \cup \{k\} \rangle$ *is a complete subgraph.*

Suppose not. Then for some $v_1, v_2 \in N(k) - K$, $v_1v_2 \notin E$. By (A1), each v_i is adjacent to at most one vertex in $K - \{k\}$. Thus there exists at least one y in K such that $v_1y, v_2y \notin E$ and $\langle k,v_1,v_2,y \rangle \simeq K_{1,3}$, a contradiction.

(A4) *If* X *and* Y *are distinct cliques in* G, $v \in X \cap Y$, *and* $|X|, |Y| \geq 4$, *then* $N[v] = X \cup Y$.

Suppose not. Let $w \in N[v] - (X \cup Y)$. By (A2), $|X \cap Y| \leq 2$. Thus $|X - Y| \geq 2$. Using (A1), it is possible to choose $x \in X \setminus Y$ such that $wx \notin E$. But then $x, w \in N(v) - Y$, contradicting (A3) because $|Y| \geq 4$.

(A5) $\delta(G) \geq 5$. *If* $\delta(G) = 5$, *then* $\omega(G) = 4$.

It is clear that for each $v \in V$, $\chi(G - v) \geq \omega(G) + 1 \geq 5$. Now if $d(v) \leq 4$, then each q-colouring of $G - v$ where $q \geq 5$, can be extended to a q-colouring of G, contradicting the assumption that G is chromatic critical. Hence $d(v) \geq 5$. The above argument also shows that if $d(v) = 5$, then $\omega(G) = 4$.

(A6) *For each* $v \in V$, *there exists a clique* K *such that* $v \in K$ *and* $|K| \geq 4$.

By (A5), for each $v \in V$, $d(v) \geq 5$. Suppose $d(v) \geq 6$. Then by Lemma 9.3, there exists $A \subseteq N(v)$ such that $\langle \{v\} \cup A \rangle \simeq K_{1,3}$ or $\langle \{v\} \cup A \rangle \simeq K_4$. Since G does not induce $K_{1,3}$, we have $v \in K$ for some clique K such that $|K| \geq 4$. Suppose $d(v) = 5$. Let $w \in N(v)$. If $d(w) \geq 6$, then by the previous argument, there exists a clique K such that $w \in K$ and $|K| = 4$ (because $\delta(G) = 5$). If $v \in K$, we are done. Otherwise, by (A3), $\langle (N(w) - K) \cup \{w\} \rangle$ is a clique of order 4 containing v. Hence we may assume that for each $w \in N(v)$, $d(w) = 5$ and v is not contained in a clique of order 4. Now since $H = \langle N(v) \rangle$ does not induce O_3 and K_3, $H = C_5 = \langle w_1,w_2,w_3,w_4,w_5 \rangle$ (see Fig.2.22). We shall next obtain a contradiction by showing that any $(\chi(G) - 1)$-colouring of $J = G - N[v]$ can be extended to a $(\chi(G) - 1)$-colouring of G. Since $\chi(G) - 1 \geq \omega(G) + 1 = 5$, and $d(w_1) = d(w_2) = 5$, we may assign the same colour α to w_1 and

w_3 in a $(\chi(G) - 1)$-colouring of J. The other vertices w_2, w_4 and w_5 can be assigned any of the other colours (except colour α) as long as

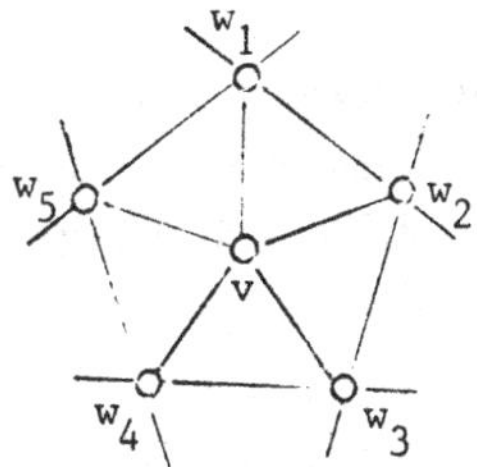

Figure 2.22

adjacent vertices do not receive the same colour. (This is possible because $\chi(G) - 1 \geq 5$ and each of the vertices w_2, w_4 and w_5 is adjacent to exactly two vertices in J.) Since H is now coloured with only four colours, the above $(\chi(G) - 1)$-colouring of G - v can be extended to a $(\chi(G) - 1)$-colouring of G, a contradiction.

(A7) For each $v \in V$, there exist distinct X, $Y \in W$ such that $v \in X \cap Y$, $N[v] \subseteq X \cup Y$, $|X| \geq 4$ and $|Y - X| \geq 2$.

By (A6), there exists a clique X such that $v \in X$ and $|X| \geq 4$. Since G is chromatic critical and $\chi(G) \geq \omega(G) + 2$, $|N(v) - X| \geq 2$. By (A3), $(N(v) - X) \cup \{v\}$ is contained in some clique Y. Thus $v \in X \cap Y$ and $N[v] \subseteq X \cup Y$. By (A2), $|X - Y| = |X| - |X \cap Y| \geq |X| - 2 \geq 2$. Clearly $|Y - X| \geq |N(v) - X| \geq 2$. Thus X, $Y \in W$.

(A8) Suppose $v \in X \cap Y$ where X and Y are distinct members of W such that $\min \{|X|, |Y|\} \geq 4$. If $v \in K$ for some other clique K, then $K \notin W$.

By (A4), $N[v] = X \cup Y$. By (A2), $|K \cap X|$, $|K \cap Y| \leq 2$. Thus $|K| \leq 3$. So X and Y are the only cliques of order at least 4 that certain v. Also $|K - X|$, $|K - Y| \leq 1$. Thus $K \notin W$.

(A7) shows that ψ is onto. We now apply (A8) to show that ψ is one-to-one. For each $v \in V$, if there exist two cliques of order at least 4 containing v, then we are done by (A8). Suppose this is not the case. By (A7), there exist cliques X, $Y \in W$ such that $v \in X \cap Y$, $N[v] \subseteq X \cup Y$, $|X| \geq 4$ and $|Y - X| \geq 2$. Since $|Y| < 4$ and $|Y - X| \geq 2$, we have $|Y| = 3$. Let $Y = \{v, z_0, z_1\}$. Suppose $K \in W$ is a clique different from X and Y containing v. By (A2), $|K \cap X|$, $|K \cap Y| < 2$. Since $K \subseteq X \cup Y$, we must have $|K \cap X| = 2 = |K \cap Y|$. Thus $|K| = 3$. Let $K = \{v, x_1, z_1\}$ where $x_1 \in X$. Since K is a clique, $x_1 z_0 \notin E$. By (A6), z_1 is contained in a clique Z of order at least 4. By (A8), v, $x_1 \notin Z$. By (A3), $N(z_1) - Z$ is a complete subgraph. Since x_1, $z_0 \in N(z_1)$, $x_1 z_0 \notin E$, and $x_1 \notin Z$, z_0 must be in Z. Since $v \in X - Z$ and vz_0, $vz_1 \in E$, by (A1), $X \cap Z = \phi$ (see Fig.2.23).

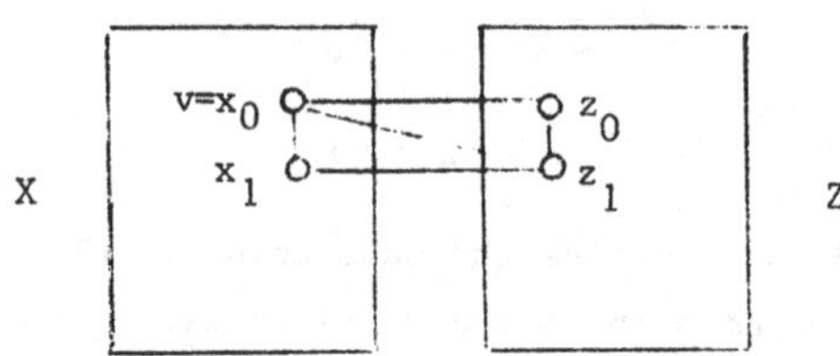

Figure 2.23

Now we show that for any $x \in X$ and any $z \in Z$, if $xz \in E$, then $N[x] \subseteq X \cup Z$ if only if $N[z] \subseteq X \cup Z$. Suppose $N[z] \subseteq X \cup Z$. If $y \in N(x)$, then by (A3), $yz \in E$ and thus $y \in X \cup Z$ which implies that $N[x] \subseteq X \cup Z$. The converse is proved by symmetry.

Next we notice that since G is chromatic critical and $\chi(G) > \omega(G) + 2$, for any $x \in X$, $|N(x) - X| > 2$ and for any $z \in Z$, $|N(z) - Z| > 2$. Thus, by (A1), if $N(x) \subseteq X \cup Z$, $|N(x) \cap Z| = 2$ and if $N(z) \subseteq X \cup Z$, $|N(z) \cap X| = 2$. We shall now define two subsets $X_0 \subseteq X$ and $Z_0 \subseteq Z$ where for each $x_i \in X_0$ and each $z_j \in Z_0$, $N[x_i]$, $N[z_j] \subseteq X \cup Z$. We first put $x_0 = v$. Clearly $x_0 \in X_0$. Then by the above result, z_0, $z_1 \in Z_0$ and thus $x_1 \in X_0$. If $z_2 \in Z$ is adjacent to x_1, then there exists $x_2 \in X$ which is adjacent to z_2 and thus $x_2 \in X_0$. We continue this process and eventually we obtain $x_k \in X_0$ such that $x_k z_k$, $x_k z_0 \in E$. Then $X_0 = \{x_0, x_1, \ldots, x_k\}$, $Z_0 = \{z_0, z_1, \ldots, z_k\}$ (see Fig.2.24).

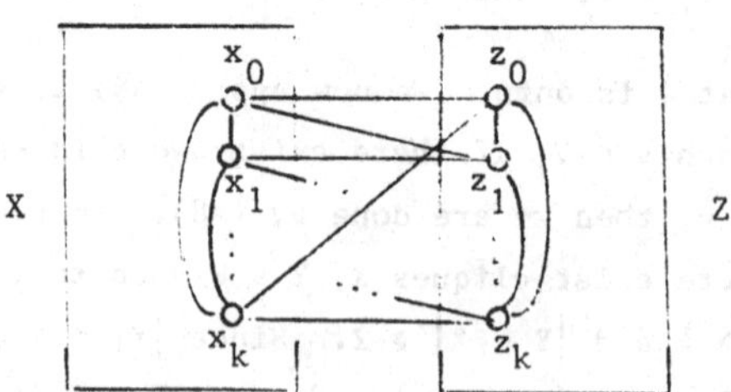

Figure 2.24

Since G is chromatic critical, there exists a $(\chi(G) - 1)$-colouring π of $H = G - (X_0 \cup Z_0)$. We will now obtain a contradiction by extending π to a $(\chi(G) - 1)$-colouring σ of G. The only elements of H, for which the elements of X_0 (resp. Z_0) are adjacent to, are in $X - X_0$ (resp. $Z - Z_0$). Let A (resp. B) be the set of colours available for colouring X_0 (resp. Z_0). Since $\chi(G) \geq \omega(G) + 2$, π is a $(\chi(G) - 1)$-colouring of H, and $|X| \leq \omega(G)$, we have $|A| \geq (\omega(G) + 1) - (|X| - |X_0|) \geq |X_0| + 1$. Similarly, $|B| \geq |Z_0| + 1$. Let $C = A \cap B = \{\gamma_0, \ldots, \gamma_{q-1}\}$, $A - C = \{\alpha_1, \ldots, \alpha_r\}$ and $B - C = \{\beta_1, \ldots, \beta_s\}$. We now extend π to σ by putting

$$\sigma(x_i) = \begin{cases} \gamma_i & \text{if } i < q \\ \alpha_{i-q+1} & \text{if } i \geq q \end{cases}$$

$$\sigma(z_i) = \begin{cases} \gamma_{i+1} & \text{if } i < q - 1 \\ \beta_{i-q+2} & \text{if } i \geq q - 1 \end{cases}$$

This contradiction completes the proof that ψ is one-to-one.

We next prove that for any $e, e' \in F$, e and e' are incident in M if and only if $\psi(e)$ and $\psi(e')$ are adjacent in G. Let $e = XY_v$, $e' = XY'_{v'}$, where $X, Y, Y' \in W$ and $v, v' \in V$. Then v and v' are in the same clique X and $\psi(e) = v$, $\psi(e') = v'$. Since ψ is one-to-one, $v \neq v'$. Thus $vv' \in E$. Conversely, suppose $vv' \in E$. Let $\psi(e) = v$ and $\psi(e') = v'$. By (A7), there exist $X, Y \in W$ such that $N[v] \subseteq X \cup Y$ and $v \in X \cap Y$. Since ψ is one-to-one, $e = XY_v$. Also, since $vv' \in E$ and $N[v] \subseteq X \cup Y$, we have $v' \in X \cup Y$. Without loss of generality, we assume that $v' \in X$. Again, since ψ is one-to-one, by (A7), there exists $Y' \in W$ such that $e' = XY'_{v'}$. Thus e and e' are incident in M.

We have so far proved that G is the line graph of M. By (A1) and the definition of F, $\mu(M) \leq 2$. The equality $\Delta(M) = \omega(G)$ follows from the fact that G is the line graph of M. Finally, if M has a 4-sided triangle, then there exist distinct $X, Y, Z \in W$ and distinct $v, v', x, y \in V$ such that $v, v' \in X \cap Y$, $x \in X \cap Z$, and $y \in Y \cap Z$. Since ψ is one-to-one, $x \notin Y$. However, the fact that x is adjacent to v, v' and y in G and $v, v', y \in Y$ contradicts (A1).

The proof of Lemma 9.7 is complete. //

To end this section, we shall prove that $M(n)$ is true for all n. This is a consequence of Theorem 9.8. We need the following definition. We call a multigraph on three vertices x, y and z such that $\mu(x,y) = s$, $\mu(x,z) = s - 1$, and $\mu(y,z) = 1$ an <u>s-triangle</u>. Thus a 2-triangle is a 4-sided triangle.

Theorem 9.8 (Kierstead [84]) <u>Let</u> M <u>be a multigraph with</u> $\Delta(M) = \Delta$ <u>and</u> $\mu(M) = \mu > 1$. <u>If</u> $q = \chi'(M) = \Delta + \mu$, <u>then</u> M <u>contains a</u> μ-<u>triangle</u>.

Proof. Suppose the theorem is false. Let M' be a multigraph such that $q = \chi'(M') = \Delta(M') + \mu(M')$ and M' does not contain a $\mu(M')$-triangle. By deleting edges from M', if necessary, we obtain a multigraph M such that

$$\chi'(M) = \chi'(M') = q = \Delta(M') + \mu(M') \geq \Delta + \mu ,$$

where $\Delta = \Delta(M)$, $\mu = \mu(M) \leq \mu(M')$, M does not contain a $\mu(M')$-triangle and M contains a critical edge $e_0 = u_0u_1$. However, by Vizing's theorem, $\chi'(M) \leq \Delta + \mu$. Hence $\chi'(M) = \Delta + \mu$, $\Delta = \Delta(M')$, $\mu = \mu(M')$ and M does not contain a μ-triangle. Let π be a (q-1)-colouring of $M - e_0$. We shall prove that π can be modified to yield a (q-1)-colouring of M, which contradicts the assumption that $\chi'(M) = q$.

We call a path $P = [u_0,u_1,\ldots,u_n]$ <u>π-acceptable</u> if for each $i > 0$, $\pi(e_i) \in \bigcup_{j<i} C'(u_j)$ where $e_i = u_iu_{i+1} \in E(M)$ and $C'(u_j) = C'_\pi(u_j)$. The notion of π-acceptable paths is a generalization of Gol'dberg's notion of "(α,β)-link chains" (Gol'dberg [77]).

The following two lemmas are the keys to the proof of Theorem 9.8.

Lemma 9.9 <u>For every</u> (q-1)-<u>colouring</u> π <u>of</u> $M - e_0$ <u>and every</u> π-<u>acceptable path</u> $P = [u_0,u_1,\ldots,u_n]$, $C'_\pi(u_i) \cap C'_\pi(u_j) = \phi$ <u>if</u> $i \neq j$.

Proof. Suppose $i < j$ and $C'(u_i) \cap C'(u_j) \neq \phi$. Let $\alpha \in C'(u_i) \cap C'(u_j)$. We prove this lemma by induction on j. If $j = 1$, then we can assign colour α to the edge e_0 to get a (q-1)-colouring of M, a contradiction.

Now suppose $j > 1$. We argue by induction on $j - i$. First suppose that $j - i = 1$. Let $\pi(e_{j-1}) \in C'(u_m)$, where $m < j - 1$. Recolour the edge e_{j-1} with colour α. Denote this new colouring of $M - e_0$ by π_1. Then $P_1 = [u_0,\ldots,u_{j-1}]$ is π_1-acceptable and $C'_{\pi_1}(u_m) \cap C'_{\pi_1}(u_{j-1}) \neq \phi$, a contradiction to induction hypothesis. Next, suppose $j - i > 1$. Let

$\beta \in C'_\pi(u_{i+1})$. By the induction hypothesis, $\beta \neq \alpha$. We now consider the $(\alpha,\beta)_\pi$-chain C having origin u_{i+1}. By the induction hypothesis, C can neither end at any of the vertices u_k for $k < i$ nor include any of the edges e_k for $k \le i$. Let π_1 be obtained from π by interchanging the colours in C. Suppose C does not end at u_i. Then $P_1 = [u_0,\ldots,u_i,u_{i+1}]$ is π_1-acceptable. But now $C_{\pi_1}(u_i) \cap C_{\pi_1}(u_{i+1}) \neq \phi$, a contradiction to the induction hypothesis. Suppose C ends at u_j. Then $P = [u_0,u_1,\ldots,u_j]$ is π_1-acceptable and $\alpha \in C'_{\pi_1}(u_{i+1}) \cap C'_{\pi_1}(u_j)$, another contradiction to the induction hypothesis. //

Lemma 9.10 Let N be the sub-multigraph of M induced by $U = \{u_0,u_1,\ldots,u_n\}$. Then $d_N(u_n) \le n(\mu - 1) + 1$.

Proof. Partition U into $\{\{u_i,u_{i+1}\} \mid i = 2j < n - 1\} \cup \{u\}$ where $\{u\} = \phi$ if n is even and $\{u\} = \{u_{n-1}\}$ if n is odd. Since N does not contain a μ-triangle and $\mu(N) \le \mu(M) = \mu$, there are at most $2\mu - 2$ edges joining u_n to each of the first $[\frac{n-1}{2}]$ parts of the partition and μ edges joining u_n to u. Thus when n is even, $d_N(u_n) \le \frac{1}{2}\, n(2\mu - 2) = n(\mu - 1)$ and when n is odd, $d_N(u_n) \le \frac{1}{2}\,(n - 1)(2\mu - 2) + \mu = n(\mu - 1) + 1$. //

Finally we prove that if P is a maximal π-acceptable path in M, then Lemma 9.9 and Lemma 9.10 cannot hold simultaneously.

Let $S = \bigcup_{i=0}^{n-1} C'(u_i)$. By Lemma 9.9, $|S| = \sum_{i=0}^{n-1} |C'(u_i)|$. But for $i \ge 2$, $|C'(u_i)| \ge (\Delta + \mu - 1) - \Delta = \mu - 1$ and for $i = 0, 1$, $|C'(u_i)| \ge (\Delta + \mu - 1) - (\Delta - 1) = \mu$. Thus $|S| \ge n(\mu - 1) + 2$. By Lemma 9.10, $d_N(u_n) \le n(\mu - 1) + 1$. Hence there exists $\gamma \in S$ such that $\pi(e) \neq \gamma$ for every edge e in N incident with u_n. Since P is maximal, $\pi(e) \neq \gamma$ for every edge $e \in E(M) - E(N)$ and incident with u_n. Consequently $\gamma \in C'(u_m) \cap C'(u_n)$ for some $m < n$, contradicting Lemma 9.9.

The proof of Theorem 9.8 is complete. //

Combining Lemma 9.1 and Theorems 9.4 and 9.8, we have

Theorem 9.11 If a graph G does not induce $K_{1,3}$ or $K_5 - e$, then $\chi(G) \le \omega(G) + 1$.

Remarks. Erdös [67] had proved that there exist positive constants c_1 and c_2 such that

(1) for every positive integer n, there is a graph G of order n satisfying

$$\chi(G) > c_1 n(\log n)^{-2}\omega(G), \text{ and}$$

(2) for every graph G of order n,

$$\chi(G) < c_2 n(\log n)^{-2}\omega(G).$$

Thus Erdös' result indicates that if a graph G induces $K_{1,3}$ or $K_5 - e$, then the actual value for $\chi(G)$ is hard to determine.

Exercise 2.9

1. Complete the proof of Lemma 9.1.
2. Prove Lemma 9.3.
3. Let G be the graph given in Fig.2.25 (due to Mycledski [55]). Prove that $\chi(G) = \omega(G) + 2$. (Note that G induces $K_{1,3}$ and $\omega(G) = 2$.)

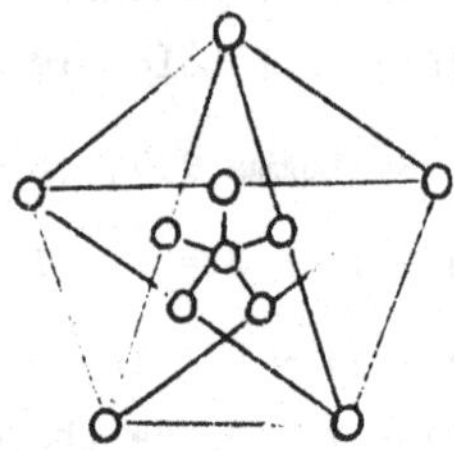

Figure 2.25

4. Let $G = C_5 + C_5$, the join of C_5 and C_5. Prove that $\chi(G) = \omega(G) + 2$ (Choudum [77]). (Note that G induces $K_5 - e$ and $\omega(G) = 4$.)

10. Applications to the reconstruction of latin squares

When a school principal is trying to construct a school timetable for his school, a good first step might seem to be to construct an outline timetable in which all Mathematics teachers are counted together, all English Language teachers are counted together, etc., all classes of each year group are counted together, and in which the preliminary division is into days rather than lessons. Having constructed an outline timetable satisfying some requirements, he might then go on to develop this outline timetable into a complete timetable.

In this section, we shall apply the theory of edge-colourings to reconstruct latin squares. This reconstruction of latin squares can be applied to the above school timetabling designs as well as to some experimental designs. The results of this section are due to Hilton [80]. We first define some terms.

A latin square L of size n is an n × n matrix on symbols 1, ..., n in which each row and each column contain each symbol exactly once.

A composition A of a positive integer n is a sequence $(a_1, \ldots, a_m)$ of positive integers such that $a_1 + a_2 + \ldots + a_m = n$. Let $P = (p_1, \ldots, p_r)$, $Q = (q_1, \ldots, q_s)$ and $S = (s_1, \ldots, s_t)$ be three compositions of n. The reduction modulo (P,Q,S) of a latin square L of size n on the symbols 1, ..., n is obtained from L by amalgamating rows $p_1 + \ldots + p_{i-1} + 1, \ldots, p_1 + \ldots + p_i$, columns $q_1 + \ldots + q_{j-1} + 1, \ldots, q_1 + \ldots + q_j$ and symbols $s_1 + \ldots + s_{k-1} + 1, \ldots, s_1 + \ldots + s_k$ for $1 \leq i \leq r$, $1 \leq j \leq s$ and $1 \leq k \leq t$. More precisely, for $1 \leq \lambda \leq r$, $1 \leq \mu \leq s$ and $1 \leq \xi \leq n$, let $x(\lambda,\mu,\xi)$ be the number of times that symbol ξ occurs in the set of cells $\{(i,j) \mid p_1 + \ldots + p_{\lambda-1} + 1 \leq i \leq p_1 + \ldots + p_\lambda, q_1 + \ldots + q_{\mu-1} + 1 \leq j \leq q_1 + \ldots + q_\mu\}$ and for $1 \leq k \leq t$, let

$$x_k(\lambda,\mu) = x(\lambda,\mu,s_1 + \ldots + s_{k-1} + 1) + \ldots + x(\lambda,u,s_1 + \ldots + s_k).$$

Then the reduction modulo (P,Q,S) of L is an r × s matrix B whose cells are filled with the symbols $\tau_1, \ldots, \tau_t$ (say) and in which cell (λ,μ) contains λ_k $x_k(\lambda,\mu)$ times.

The following is an example of the reduction modulo (P,Q,S) of L.

Example. Let L be the latin square of size 12 given in Fig.2.26 in which $\pi = 10$, $\infty = 11$ and $e = 12$. Let $r = 3$, $s = 4$ and $t = 4$, and let $P = (p_1,p_2,p_3) = (4,4,4)$, $Q = (q_1,q_2,q_3,q_4) = (5,3,3,1)$ and $S = (s_1,s_2,s_3,s_4) = (4,3,3,2)$. Let I be the composition consisting of a sequence of the appropriate length of 1's.

e	1	2	3	4	5	6	7	8	9	π	∞
∞	π	e	9	1	7	2	8	3	6	4	5
π	e	∞	1	9	2	8	3	4	7	5	6
9	∞	π	8	e	1	7	2	5	3	6	4
5	6	7	π	8	e	4	1	9	2	∞	3
1	2	3	4	5	6	9	∞	7	8	e	π
4	9	1	5	2	π	3	6	∞	e	7	8
2	3	4	e	6	8	5	π	1	∞	9	7
6	7	8	∞	π	3	e	5	2	4	1	9
7	8	5	6	3	4	∞	9	π	1	2	e
8	4	9	2	7	∞	π	e	6	5	3	1
3	5	6	7	∞	9	1	4	e	π	8	2

Figure 2.26

1	1	1	2	3	1	2	2	3	3	4	4
4	8	9	9	9	2	3	5	4	5	5	5
π	π	π	∞	∞	6	7	7	6	6	7	6
∞	e	e	e	e	7	8	8	8	9	π	π
1	1	2	2	2	1	3	4	1	2	7	3
3	3	4	4	4	5	6	6	7	8	9	7
5	5	5	6	6	8	9	π	9	∞	∞	8
7	8	9	π	e	π	∞	e	∞	e	e	π
2	3	3	4	5	1	3	4	1	1	2	1
5	6	6	6	7	4	5	9	2	3	4	2
7	7	7	8	8	9	π	∞	5	6	8	9
8	9	π	∞	∞	∞	e	e	π	π	e	e

Figure 2.27

The reduction of L modulo (P,Q,I) is given in Fig.2.27. In this figure, there is not intended to be any significance in the way the symbols are arranged in each cell.

Finally, we obtain the reduction modulo (P,Q,S) of L by replacing 1, 2, 3 and 4 by $\alpha = \tau_1$; 5, 6 and 7 by $\beta = \tau_2$; 8, 9 and π by $\delta = \tau_3$; and ∞ and e by $\varepsilon = \tau_4$ (see Fig.2.28).

α α δ ε	α δ δ ε	α δ δ ε	α δ ε ε	α δ ε ε	α α β β	α α β δ	α β β δ	α α β δ	α β β δ	α β β δ	α β β δ
α α β β	α α β δ	α α β δ	α α β δ	α α β ε	α β δ δ	α β δ ε	α β δ ε	α β δ ε	α δ ε ε	β δ ε ε	α β δ ε
α β β δ	α β β δ	α β β δ	α β δ ε	β β δ ε	α α δ ε	α β δ ε	α δ ε ε	α α β δ	α α β δ	α α δ ε	α α δ ε

Figure 2.28

We now define an outline rectangle. Let C be an $r \times s$ matrix filled with t symbols $\tau_1,\ldots,\tau_t$ in which each cell may be occupied by more than one symbol and in which each symbol may occur more than once. For $1 \leq i \leq r$, $1 \leq j \leq s$ and $1 \leq k \leq t$, let ρ_i be the number, including repetitions, of symbols which occur in row i, let c_j be the number, including repetitions, of symbols which occur in column j and let σ_k be the number of times τ_k appears in C. Then C is called an <u>outline rectangle</u> if, for some integer n, the following properties are obeyed for each i, j, k such that $1 \leq i \leq r$, $1 \leq j \leq s$ and $1 \leq k \leq t$:

(i) n divides each ρ_i, c_j and σ_k;

(ii) cell (i,j) contains $(1/n^2)\rho_i c_j$ symbols (including repetitions);

(iii) the number of times τ_k appears in row i is $(1/n^2)\rho_i \sigma_k$;

(iv) the number of times τ_k appears in column j is $(1/n^2)c_j\sigma_k$.

From the above definitions, the reduction modulo (P,Q,S) of a latin square L is an outline rectangle and has the further properties

(v) $(p_1,\ldots,p_r) = (\frac{\rho_1}{n},\ldots,\frac{\rho_r}{n})$;

(vi) $(q_1,\ldots,q_s) = (\frac{c_1}{n},\ldots,\frac{c_s}{n})$;

(vii) $(s_1,\ldots,s_t) = (\frac{\sigma_1}{n},\ldots,\frac{\sigma_t}{n})$;

(viii) $\sum_{i=1}^{r}\rho_i = \sum_{j=1}^{s}c_j = \sum_{k=1}^{t}\sigma_k = n^2$.

We shall now show that any outline rectangle could have been formed from some latin square by reduction modulo (P,Q,S) for some suitable compositions P, Q and S. The main tool is a theorem of de Werra on balanced edge-colourings of a multigraph M.

Suppose π is a k-edge-colouring (not necessarily proper) of a loopless multigraph M. For each $v \in V(M)$, let $E_i(v)$ be the set of all edges incident with v each of which is coloured with colour i, and for each $u, v \in V(M)$, $u \neq v$, let $E_i(u,v)$ be the set of all edges joining u and v each of which is coloured with colour i. Then π is said to be <u>equitable</u> if, for all $v \in V(M)$,

(a) $$\max_{1\leq i,j\leq k} \left||E_i(v)| - |E_j(v)|\right| \leq 1$$

and π is said to be <u>balanced</u> if, in addition, for all $u,v \in V(M)$, $u \neq v$,

(b) $$\max_{1\leq i,j\leq k} \left||E_i(u,v)| - |E_j(u,v)|\right| \leq 1.$$

Thus an edge-colouring of M is balanced if the colours occur as uniformly as possible at each vertex of M and if the colours are shared out as uniformly as possible on each multiple edge of M.

We shall apply the following theorem due to de Werra [71,75a,75b]. The proof given here is due to Andersen and Hilton [79].

Theorem 10.1 <u>For each integer</u> $k \geq 1$, <u>any finite bipartite multigraph</u> M <u>has a balanced edge-colouring with</u> k <u>colours</u>.

Proof. Colour the edges of the multigraph in such a way that (b) is satisfied; condition (b) only affects each multiple edge by itself, so this is clearly possible. We then modify the colouring to make (a) be satisfied without violating (b). Suppose that at some vertex v,

$$\max_{1 \leq i < j \leq k} \left| |E_i(v)| - |E_j(v)| \right| > 1.$$

We may suppose that this maximum is attained for colours 1 and 2 and that $|E_1(v)| > |E_2(v)| + 1$. Let P be a maximal chain $v = v_0, e_1, v_1, e_2, v_2, \ldots, e_h, v_h$ (where e_i is an edge joining v_{i-1} to v_i and $e_i \neq e_j$ if $i \neq j$) such that

(I) e_1 is coloured with colour 1,

(II) $e_1, \ldots, e_h$ are coloured alternately with colours 1 and 2,

(III) $|E_1(v_i, v_{i+1})| = |E_2(v_i, v_{i+1})| + 1$ if i is even, and

$|E_2(v_i, v_{i+1})| = |E_1(v_i, v_{i+1})| + 1$ if i is odd,

(IV) P uses only one edge from each multiple edge.

(Note that the same vertex may occur several times in P.)
Then $h \neq 0$ because v has some neighbour v_1 for which $|E_1(v,v_1)| = |E_2(v,v_1)| + 1$, since $|E_1(v)| > |E_2(v)| + 1$. Also $v_h \neq v_0$, because if $v_j = v_0$, then j is even as the multigraph is bipartite, so when j edges have been traversed, both colours have occurred the same number of times in total on the multiple edges incident with v_0 used so far and so the chain can be continued since $|E_1(v)| > |E_2(v)| + 1$.

Interchanging the two colours 1 and 2 on the chain P clearly does not violate (b), it reduces the number of pairs of colours for which $\left| |E_i(v)| - |E_j(v)| \right|$ was maximal by at least one and it does not affect

$$\max_{1 \leq i < j \leq k} \left| |E_i(v_t)| - |E_j(v_t)| \right|$$

if $0 < t < h$.

If h is odd, then e_h is coloured with colour 1, so the maximality of P implies that the number of multiple edges (v_h, x) on v_h for which

$$|E_1(v_h, x)| = |E_2(v_h, x)| + 1$$

exceeds the number of multiple edges on v_h for which

$$|E_2(v_h,x)| = |E_1(v_h,x)| + 1$$

by at least one, so colour 1 occurs at least once more than colour 2 at v_h. Thus

$$\max_{1\leq i<j\leq k} \big||E_i(v_h)| - |E_j(v_h)|\big|$$

will not be increased. A similar statement is true if h is even.

Repeated application of the argument then proves Theorem 10.1. //

We now apply Theorem 10.1 to prove the main theorem of this section.

Theorem 10.2 (Hilton [80]) _To each outline rectangle_ C _there is a latin square_ L _and there are compositions_ P, Q _and_ S _such that_ C _is the reduction of_ L _modulo_ (P,Q,S).

Proof. First we observe that if m is the number of entries in C, then

$$m = \sum_{i=1}^{r} \rho_i = \sum_{j=1}^{s} c_j = \sum_{i=1}^{r} \sum_{j=1}^{s} \frac{1}{n^2} \rho_i c_j = \frac{1}{n^2} \sum_{i=1}^{r} \rho_i (\sum_{j=1}^{s} c_j) = \frac{m^2}{n^2},$$

so $m = n^2$.

We next observe that the outline rectangle can be represented as a family of triples (x,y,z) where each occurrence of each symbol in each cell of C corresponds to exactly one triple, the first coordinate denoting the row the cell lies in, the second coordinate denoting the column the cells lies in and the third coordinate denoting the index of the symbol τ_k that the cell contains. Thus if cell (i,j) contains the symbol τ_k we obtain the triple (i,j,k). There are therefore n^2 triples, counting repetitions. The conditions (ii), (iii) and (iv) now take on the move symmetrical form:

(ii)' (i,j) occurs as the first pair in $(1/n^2)\rho_i c_j$ triples;

(iii)' (i,k) occurs as the first and last entries in $(1/n^2)\rho_i \sigma_k$ triples.

(iv)' (j,k) occurs as the last pair in $(1/n^2)c_j\sigma_k$ triples.

Because of this symmetry we may, without loss of generality, confine the explanation to the case that $r < n$ and show that C can be obtained from an $(r + 1) \times s$ outline rectangle C' by amalgamating the cells of two rows (so that any pair of cells in these two rows which are in the same column are identified) or, in other words, by reduction modulo (P^*,I,I), where P^* is a composition with one term 2, the rest all 1's. Repeated application of this argument first on the rows, then on the columns and finally on the symbols will show that C can be obtained from an $n \times n$ outline rectangle on n symbols, i.e. a latin square, by reduction modulo (P,Q,S) for some compositions P, Q and S.

Since $r < n$, n divides $\rho_1, \ldots, \rho_r$ and $\sum_{i=1}^{r} \rho_i/n = n$, there is at least one i for which $\rho_i/n > 1$. We may assume, without loss of generality, that $\rho_r/n \geq 2$. We wish to form an outline rectangle C' by splitting the last row of C into two new rows. We construct a bipartite multigraph M with vertex classes $\{\gamma_1,\ldots,\gamma_s\}$ and $\{\tau_1,\ldots,\tau_t\}$ where the vertex γ_j is joined to the vertex τ_k by y edges if and only if the symbol τ_k occurs y times in cell (r,j) of C. Then the valency of γ_j is the number, including repetitions, of symbols in the cell, namely $(1/n^2)\rho_r c_j$, and the valency of the vertex τ_k is the number of times τ_k occurs in row r of C, namely $(1/n^2)\rho_r\sigma_k$.

We now give M a balanced edge-colouring with ρ_r/n colours. Let E_1 be the set of those edges coloured with colour 1. Then each vertex γ_j is incident with exactly

$$\frac{n}{\rho_r}\left(\frac{1}{n^2}\rho_r c_j\right) = \frac{1}{n}c_j$$

edges receiving colour 1. Now split row r of C into two rows r' (to be row $r + 1$ of C') and r" (to be row r of C') by placing a symbol τ_k in cell (r',j) x times if and only if there are x edges of colour 1 joining the vertices γ_j and τ_k and by placing τ_k in cell (r'',j) y times if and only if there are y edges of colours different from 1 joining the vertices γ_j and τ_k.

We now check that C' is an outline rectangle . Let $\rho_i' = \rho_i$ $(1 \leq i < r)$, $\rho_r' = \rho_r - n$ and $\rho_{r+1}' = n$. Let $c_j' = c_j$ $(1 \leq j \leq s)$ and $\tau_k' = \tau_k$

$(1 \leq k \leq t)$. Then clearly n divides each of ρ_i', c_j' and σ_k'. Cells (r,j) and $(r+1,j)$ of C' contain, respectively,

$$\frac{1}{n^2} c_j(\rho_r - n) = \frac{1}{n^2} c_j'\rho_r' \quad \text{and} \quad \frac{1}{n} c_j = \frac{1}{n^2} \rho_{r+1}'c_j'$$

symbols, including repetitions. Each symbol τ_k occurs

$$\frac{1}{n^2} (\rho_1 - n)\sigma_k = \frac{1}{n^2} \rho_r'\sigma_k'$$

times in row r of C' and

$$\frac{1}{n} \sigma_k = \frac{1}{n^2} \rho_{r+1}'\sigma_k'$$

times in row $r + 1$. Thus in these cells and rows conditions (ii), (iii) and (iv) applied to C' are satisfied and also they are clearly satisfied in all columns and all other cells and rows.

This proves Theorem 10.2. //

Exercise 2.10

1. Show that if a bipartite multigraph M has a balanced edge-colouring π, then M has a balanced edge-colouring π^* such that the colour classes $E_1, \ldots, E_k$ of π^* can be made in such way that $|E_1| \leq |E_2| \leq \ldots \leq |E_k| \leq |E_1| + 1$ (Bollobás [79;p.63,Ex.9]).
2. Let C be the outline rectangle given by Fig.2.28. Using the balanced edge-colouring technique of the proof of Theorem 10.2, reconstruct a latin square L so that C is the reduction of L modulo (P,Q,S) for some compositions P, Q and S.

11. Concluding remarks

To conclude this chapter, we shall now briefly mention some other interesting and important results concerning edge-colourings which we have not been able to discuss in detail due to lack of space.

1. Snarks.

Due to the fact that the Four-Colour Conjecture is equivalent to the assertion that every bridgeless cubic planar graph is 3-edge-colourable, much attention has been paid to the the search for bridgeless cubic graphs which are not 3-edge-colourable. Since such graphs are difficult to find, Gardner [76] christened them snarks after Lewis Carroll's "The Hunting of the Snark". (Even after the Four-colour Conjecture has been proved, many people are still searching for non-planar snarks.)

Suppose G is a cubic graph which contains a triangle K_3 and $\chi'(G')$ $= 4$. Let G^* be the graph obtained from G by contracting K_3 into a vertex. Then $\chi'(G^*) = 3$ implies that $\chi'(G) = 3$, a contradiction. Hence, in order to avoid trivial cases, we assume that a snark does not contain triangles. For similar reasons, we also assume that a snark does not contain quadrilaterals and does not contain three or two edges whose deletion results in a disconnected graph each of whose component is nontrivial. Hence a snark can be precisely defined as follows : A snark G is a cubic graph of girth at least 5 with $\chi'(G) = 4$, which is cyclically-4-edge-connected (i.e. G does not contain three or fewer edges whose deletion results in a disconnected graph, each of whose component is nontrivial).

The smallest snark is the Petersen graph found in 1898. The second snark (of order 18) was found by Blanuša in 1946. The third snark (of order 20) was found by Descartes in 1948. The fourth snark (of order 50) was found by Szekeres in 1973.

Two infinite families of snarks were found by Isaacs in 1975. The first family includes all the four known snarks. The second family has been discovered independently by Grinberg in 1972, but never published. Isaacs also discovered a snark of order 30, which does not fit into either of the two infinite families. In some unpublished work done in 1976, F. Loupekhine also constructed an infinite family of snarks.

The fourth infinite family of snarks of even order (see §4) was found by Gol'dberg in 1979. Further snarks have been found by U. A.

Celmins and E. R. Swart (University of Waterloo Research Report CORR 79-18).

The following is a long standing conjecture concerning snarks.

Tutte's Conjecture (Tutte [69]) *Every snark contains a subgraph homeomorphic to the Petersen graph.*

2. Edge-colourings and groups

Cameron [75a, 75b] studied relations between the colour classes of an (n-1)-edge-colouring of K_n, n even, and some permutation groups of the vertices of K_n (see Ex.2.1(7) and 2.1(9)). Further results concerning edge-colourings and groups can also be found in Biggs [72] and Cameron [76].

3. The Total Colouring Conjecture

The *total chromatic number* $\chi_2(G)$ of a graph G is the minimum number of colours required to colour the elements (vertices and edges) of G in such a way that two adjacent, as well as two incident, elements of G receive different colours. It is obvious that for any graph G, $\chi_2(G) \geq \Delta(G) + 1$. In 1965, M. Behzad made the following conjecture.

The Total Colouring Conjecture *For any graph* G, $\chi_2(G) \leq \Delta(G) + 2$.

It is easy to show that $\chi_2(K_{3,3}) = 5$. Thus if the Total Colouring Conjecture is true, then it is best possible. Rosenfeld [71] proved this conjecture for bipartite, complete tripartite and complete balanced n-partite graphs, and also for all graphs G with $\Delta(G) \leq 3$. However, this long standing conjecture still remains open. For a survey of results on this conjecture, see Behzad [71].

In a recent paper by Bollobás and Harris [85], it is proved that if $\Delta(G)$ is large enough, then $\chi_2(G) < c\Delta(G)$ where $11/6 < c < 2$.

4. Extensions of partial edge-colourings

A *partial edge-colouring* of a graph G is an edge-colouring of some subgraph G' of G. Andersen and Hilton [80] showed that any partial edge-colouring of K_n with at most 2n - 1 colours can be extended to an edge-colouring of K_{2n} with 2n - 1 colours. They also showed that any partial edge-colouring of K_n with 2n + 1 colours can be extended to an edge-colouring of K_{2n+1} with 2n + 1 colours. The idea of extending a partial edge-colouring has been further explored in Chetwynd and Hilton [84a].

5. Feasible sequence for a graph

A non-increasing sequence $F = (f_1, f_2, \ldots, f_k)$ of non-negative integers is feasible for a graph G if there is a k-edge-colouring π of G such that the colour classes $E_1, E_2, \ldots, E_k$ of π are such that $|E_i| = f_i$, $i = 1, 2, \ldots, k$.

The following two theorems are very useful in constructing chromatic index critical graphs, see for instance, Plantholt [-a].

Theorem 11.1 (de Werra [71b], McDiarmid [72]) *If the non-increasing sequence* $F = (f_1, f_2, \ldots, f_k)$ *is feasible for a graph* G, *then so is any sequence* $F' = (f'_1, f'_2, \ldots, f'_k)$ *such that*

$$\sum_{i=1}^{k} f'_i = \sum_{i=1}^{k} f_i \quad \text{and} \quad \sum_{i=1}^{j} f'_i \leq \sum_{i=1}^{j} f_i \quad \text{for} \quad j = 1, 2, \ldots, k-1.$$

Theorem 11.2 (Folkman and Fulkerson [69]) *Suppose* $F = (f_1, f_2, \ldots, f_n)$ *is an increasing sequence of non-negative integers. Let* $f^* = \max\{i \mid f_i \geq 1\}$ *and let* B *be a bipartite graph of size* $\sum_{f=1}^{\infty} f^*$. *Then* F *is feasible for* B *if and only if*

$$e(B - X) \geq \sum_{f=|X|+1}^{\infty} f^* \quad \text{for all} \quad X \subset V(B).$$

6. The circumference of critical graphs

Vizing [65b], using his Adjacency Lemma, proved that if G is a Δ-critical graph, then G contains a circuit whose length is at least $\Delta + 1$. Fiorini [75c] proved that if G is a Δ-critical graph of order n whose minimum valency is δ, then G contains a circuit whose length is at least

$$2\frac{\log((n-1)(\Delta-2)/\delta)}{\log(\Delta-1)} .$$

7. The girth of critical graphs

Fiorini [76] proved the following theorems concerning the girth of chromatic-index critical graphs.

Theorem 11.3 *For any integers* $\Delta \geq 3$ *and* $\gamma \geq 3$, *there exists a* Δ-*critical graph of girth* γ.

Theorem 11.4 *Let* G *be a* Δ-*critical graph of order* n *and girth* γ.

(i) *If* $\gamma = 3$, *then* $n \geq \Delta + 1$ *if* Δ *is even*; and $n \geq \Delta + 2$ *if* Δ *is odd*.

(ii) *If* $\gamma = 4$, *then* $n \geq 2\Delta + 1$.

Moreover, there exist critical graphs which attain these bounds.

8. Uniquely colourable graphs

We refer to Ex.2.2(7) for the definition of a uniquely k-colourable graph. It was conjectured in Fiorini [75a] and Wilson [75] that if $k \geq 4$, then the only uniquely k-colourable graph is the star S_{k+1}. This conjecture has been proved by Thomason [78]. For $k = 3$, the following two conjectures are still open.

Conjecture 1 (Fiorini and Wilson [77]) *Every uniquely* 3-*colourable cubic planar graph contains a triangle*.

Conjecture 2 (Greenwell and Kronk [73]) *If* G *is a cubic graph with exactly three Hamilton cycles, then* G *is uniquely* 3-*colourable*.

A bibliography on edge-colourings of graphs can be found in Fiorini and Wilson [77]. The list of papers published after 1977 appearing in the references of this chapter can be served as a complement to the bibliography compiled by Fiorini and Wilson.

REFERENCES

Y. Alavi and M. Behzad, Complementary graphs and edge chromatic numbers, Siam J. Appl. Math., 20 (1971), 161-163.

L. D. Andersen, On edge-colourings of graphs, Math. Scand., 40, no.2, (1977), 161-175.

L. D. Andersen and A. J. W. Hilton, Generalized latin rectangles, Proceedings of the one-day conference on combinatorics at the Open University, Pitman, London (1979), 1-17.

———, Generalized latin rectangles II : Embedding, Discrete Math., 31 (1980), 235-260.

M. Behzad, The total chromatic number of a graph : A survey, Combinatorial Mathematics and its Applications (Proc. Conf. Oxford, 1969), Academic Press, London (1971), 1-8.

M. Behzad, G. Chartrand and J. K. Cooper, The colour numbers of complete graphs, J. London Math. Soc., 42 (1967), 226-228.

L. W. Beineke, Derived graphs and digraphs, Beitrage zur Graphen theorie (Eds., H. Sachs, H. Voss and H. Walther), Teubner, Leipzig (1968), 17-33.

L. W. Beineke and S. Fiorini, On small graphs critical with respect to edge-colourings, Discrete Math., 16 (1976), 109-121.

L. W. Beineke and R. J. Wilson, On the edge-chromatic number of a graph, Discrete Math., 5 (1973), 15-20.

C. Berge, Graphs and Hypergraphs, North-Holland Mathematical Library, Vol.6 (1973).

N. L. Biggs, Pictures, Combinatorics (Eds., D. J. A. Welsh and D. R. Woodall), Inst. Math. Apl., Southend-on-Sea (1972), 1-17.

———, Three remarkable graphs, Canad. J. Math., 25 (1973), 397-411.

D. Blanuša, The problem of four colours, Hrvatsko Prirodoslovno Društvo Glasnik Mat. - Fiz. Astr., Ser.II, 1 (1946), 31-42.

B. Bollobás, Graph Theory - An Introductory Course, Graduate Text in Mathematics 63, Springer-Verlag, 1979.

B. Bollobás and A. J. Harris, List-colourings of graphs, Graphs and Combinatorics 1 (1985), 115-127.

J. Bosák, Chromatic index of finite and infinite graphs, Czech. Math. J., 22 (1972), 272-290.

I. Broere and C. M. Mynhardt, Remarks on the Critical Graph Conjecture, Discrete Math., 26 (1979), 209-212.

R. A. Brualdi, The chromatic index of the graph of the assignment polytope, Ann. Discrete Math., 3 (1978), 49-53.

P. J. Cameron, On groups of degree n and n - 1, and highly symmetric edge-colourings, J. London Math. Soc., (2) 9 (1975), 385-391.

———, Minimal edge-colourings of complete graphs, J. London Math. Soc., (2) 11, no.3 (1975), 337-346.

———, Parallelisms of Complete Designs, London Math. Soc. Lecture Notes, No.23, Cambridge Univ. Press (1976)

F. Castagna and G. Prins, Every generalized Petersen graph has a Tait colouring, Pacific J. Math., 40 (1972), 53-58.

A. G. Chetwynd and A. J. W. Hilton, Partial edge-colourings of complete graphs or of graphs which are nearly complete, Graph Theory and Combinatorics (Ed., B. Bollobàs), Academic Press, London (1984), 81-97.

———, The chromatic index of graphs of even order with many edges, J. Graph Theory 8 (1984), 463-470.

———, Regular graphs of high degree are 1-factorizable, Proc. London Math. Soc., 50 (3) (1985), 193-206.

———, The chromatic index of graphs with at most four vertices of maximum degree, Proc. of the 15th Southeastern Conf. on Combinatorics, Graph Theory and Computing, Congre. Numer., 43 (to appear).

———, The chromatic index of graphs with large maximum degree, where the number of vertices of maximum degree is relatively small (preprint).

———, The edge chromatic class of regular graphs of degree 4 and their complements (preprint).

A. G. Chetwynd and R. J. Wilson, Snarks and supersnarks, The Theory of Applications of Graphs (Ed., G. Chartrand), John Wiley and Sons, Inc., (1981), 215-241.

———, The rise and fall of the Critical Graph Conjecture, J. Graph Theory 7 (1983), 153-157.

A. G. Chetwynd and H. P. Yap, Chromatic index critical graphs of order 9, Discrete Math., 47 (1983), 23-33.

S. A. Choudum, Chromatic bounds for a class of graphs, Quart. J. Math. Oxford (2), 28 (1977), 257-270.

D. de Werra, Balanced schedules, Infor Journal 9 (1971), 230-237.

———, Equitable colorations of graphs, Rev. France Rech. Oper., 5 (1971), 3-8.

———, A few remarks on chromatic scheduling, Combinatorial Programming : Methods and Applications (Ed., B. Roy), D. Reidel, Dordrecht, Holland (1975), 337-342.

———, On a particular conference scheduling problem, Infor Journal 13 (1975), 308-315.

B. Descartes, Network colourings, Math. Gazette 32 (1948), 67-69.

G. A. Dirac, A property of 4-chromatic graphs and some remarks on critical graphs, J. London Math. Soc., 27 (1952), 85-92.

A. Ehrenfeucht, V. Faber and H. A. Kierstead, A new method of proving theorems on chromatic index, Discrete Math., 52 (1984), 159-164.

P. Erdös, Some remarks on chromatic graphs, Colloquium Math., 16 (1967), 253-256.

P. Erdös and R. J. Wilson, On the chromatic index of almost all graphs, J. Combin. Theory, Ser.B, 23 (1977), 255-257.

A. Eshrat, An efficient algorithm for colouring the edges of a graph with $\Delta + 1$ colours, Discrete Mathematical Analysis and Combinatorial Computation, Frederiction, N. B. (1980), 108-132.

R. J. Faudree and J. Sheehan, Regular graphs and edge chromatic number, Discrete Math., 48 (1984), 197-204.

M. A. Fiol, 3-grafos criticos, Doctoral dissertation, Barcelona University, Spain (1980) (quoted from Chetwynd and Wilson [83]).

S. Fiorini, On the chromatic index of a graph, III : Uniquely edge-colouring graphs, Quart. J. Math., Oxford (3), 26 (1975),129-140.

———, On the chromatic index of outerplanar graphs, J. Combin. Theory, Ser.B, 18 (1975), 35-38.

———, Some remarks on a paper by Vizing on critical graphs, Math. Proc. Camb. Phil. Soc., 77 (1975), 475-483.

———, On the girth of graphs critical with respect to edge-colourings, Bull. London Math. Soc., 8 (1976), 81-86.

———, Counterexamples to two conjectures of Hilton, J. Graph Theory 2 (1978), 261-264.

S. Fiorini and R. J. Wilson, On the chromatic index of a graph, I, Cahiers centre Études Rech. Opèr., 15 (1973), 253-262.

———, On the chromatic index of a graph, II. Combinatorics (Eds., T. P. McDonough and V. C. Mavron), Cambridge Univ. Press, (1974), 37-51.

———, Edge-colourings of graphs: some applications, Proceedings of the Fifth British Combinatorial Conference, 1975, Utilitas Math., Winnipeg (1976), 193-202.

———, Edge-colourings of Graphs, Research Notes in Mathematics, Vol.16, Pitman, London (1977).

J. Folkman and D. R. Fulkerson, Edge colourings in bipartite graphs, Combinatorial Mathematics and Its Applications (Eds., R. C. Bose and T. A. Dowling), University of North Carolina Press, Chapel Hill (1969), 561-577.

M. Gardner, Mathematical games, Scientific American 234 (1976), 126-130.

M. K. Gol'dberg, Remarks on the chromatic class of multigraphs (Russian), Vycislitel'naya Matematika i Vycislitel'naya Tekhnika 5 (1974), 128-130.

———, Structure of multigraphs with constraints on the chromatic class (Russian), Diskret. Analiz 30 (1977), 3-12.

———, Critical graphs with an even number of vertices (Russian), Bull. Acad. Sci. Georgia, SSR, 94 (1979), 25-27.

———, On graphs of (maximum) degree 3 with chromatic index 4 (Russian), Bull. Acad. Sci. Georgia, SSR, 93 (1979), 29-31.

———, Construction of class 2 graphs with maximum vertex degree 3, J. Combin. Theory, Ser.B, 31 (1981), 282-291.

———, Edge-colouring of multigraphs : Recolouring technique, J. Graph Theory 8 (1984), 123-137.

D. Greenwell and H. Kronk, Uniquely line-colorable graphs, Canad. Math. Bull., 16 (1973), 525-529.

R. P. Gupta, The chromatic index and the degree of a graph, Notices Amer. Math. Soc., 13 (1966), 719.

———, An edge-coloration theorem for bipartite graphs with applications, Discrete Math., 23 (1978), 229-233.

G. Hajós, Über eine Konstruktion nicht n-färbarer Graphen, Wiss. Z. Martin-Luther Univ. Halle-Wittenberg Math. Nat. Reihe, 10 (1961), 116-117.

A. J. W. Hilton, On Vizing's upper bound for the chromatic index of a graph, Cahiers Certre Études Rech. Opér., 17 (1975), 225-233.

———, Definitions of criticality with respect to edge-colouring, J. Graph Theory 1 (1977), 61-68.

———, The reconstruction of latin squares with applications to school timetabling and to experimental design, Math. Programming Study 13 (1980), 68-77.

———, Canonical edge-colourings of locally finite graphs, Combinatorica 2, no.1, (1982), 37-51.

A. J. Hilton and D. de Werra, Sufficient conditions for balanced and for equitable edge-colourings (preprint).

A. J. W. Hilton and C. A. Roger, Edge-colouring regular bipartite graphs, Graph Theory, North-Holland Math. Stud., 62 (1981), 139-158.

I. Holyer, The NP-completeness of edge-colouring, Siam J. Comput., 10 (1981), 718-720.

Huang Ji Hua, Uniquely 3-edge-colourable graphs (Chinese), Acta Math. Appl. Sinica 7 (1984), 285-292.

R. Isaacs, Infinite families of non-trivial trivalent graphs which are not Tait colorable, Amer. Math. Monthly 82 (1975), 221-239.

F. Jaeger, Sur l'indice chromatique du graphe représentatif des arêtes d'un graphs regulier, C. R. Acad. Sci., Paris, Ser.A, 277 (1973), 237-239.

F. Jaeger and H. Shank, On the edge-colouring problem for a class of 4-regular maps, J. Graph Theory 5, no.3, (1981), 269-275.

I. T. Jakobsen, Some remarks on the chromatic index of a graph, Arch. Math. (Basel), 24 (1973), 440-448.

———, On critical graphs with chromatic index 4, Discrete Math., 9 (1974), 265-276.

———, On graphs critical with respect to edge-colouring, Infinite and Finite Sets, Keszthely, 1973, North-Holland (1975), 927-934.

M. Javdekar, Note on Choudum's "Chromatic bound for a class of graphs", J. Graph Theory 4 (1980), 265-268.

E. I. Johnson, A proof of four-coloring the edges of a cubic graph, Oper. Res. Centre, Univ. California (1963) = Amer. Math. Monthly 73 (1966), 52-55.

H. A. Kierstead, On the chromatic index of multigraphs without large triangles, J. Combin. Theory, Ser.B, 36 (1984), 156-160.

H. A. Kierstead and J. H. Schmerl, Some applications of Vizing's theorem to vertex colorings of graphs, Discrete Math., 45 (1983), 277-285.

D. König, Über Graphen und ihre Anwendung auf Determinanten-theorie und Mengenlehre, Math. Ann., 77 (1916), 453-465.

R. Laskar and W. Hare, Chromatic numbers of certain graphs, J. London Math. Soc., (2) 4 (1971), 489-492.

C. McDiarmid, The solution to a time-tabling problem, J. Inst. Maths. Applics., 9 (1972), 23-34.

L. S. Mel'nikov, The chromatic class and the location of a graph on a closed surface, Mat. Zametki 7 (1970), 671-681; Math. Notes 7 (1970), 405-411.

———, Some topological classifications of graphs, Recent Advances in Graph Theory (Ed., M. Fiedler), Academia, Prague (1975), 365-383.

J. Mycielski, Sur le coloriage des graphes, Colloquium Math., 3 (1955), 161-162.

B. H. Neumann, An embedding theorem for algebraic systems, Proc. London Math. Soc., (3) 4 (1954), 138-153.

O. Ore, The Four-Colour Problem, Academic Press, New York (1967).

E. T. Parker, Edge coloring numbers of some regular graphs, Proc. Amer. Math. Soc., 37 (1973), 423-424.

M. Plantholt, The chromatic index of graphs with a spanning star, J. Graph Theory 5 (1981), 5-13.

———, The chromatic index of graphs with large maximum degree, Discrete Math., 47 (1983), 91-96.

———, A generalized construction of chromatic index critical graphs from bipartite graphs (preprint).

M. Preissmann, Snarks of order 18, Discrete Math., 42, no.1, (1982), 125-126.

A. Rosa and W. D. Wallis, Premature sets of 1-factors or how not to schedule round-robin touraments, Discrete Applied Math., 4 (1982), 291-297.

M. Rosenfeld, On the total coloring of certain graphs, Israel J. Math., 9 (1971), 396-402.

C. E. Shannon, A theorem on coloring the lines of a network, J. Math. Phys., 28 (1949), 148-151.

Sun Hui Cheng, Ni Jin and Chen Guan Di, The criticality of the chromatic index of a graph (Chinese), Nanjing Daxue Xuebao, no.4, (1982), 832-836.

G. Szekeres, Polyhedral decompositions of cubic graphs, Bull. Austral. Math. Soc., 8 (1973), 367-387.

P. G. Tait, Remarks on the colouring of maps, Proc. Royal Soc. Edin., 10 (1880), 501-503, 729.

———, Note on a theorem in the geometry of position, Trans. Royal Soc. Edin., 29 (1898), 270-285.

A. Thomason, On uniquely colourable graphs of maximum valency at least four, Ann. Discret Math., 3 (1978), 259-268.

W. T. Tutte, A geometrical version of the four color problem, Combinatorial Mathematics and its Applications (Eds., R. C. Bose and T. A. Dowling), University of North Carolina Press (1969).

V. G. Vizing, On an estimate of the chromatic class of a p-graph (Russian), Diskret. Analiz 3 (1964), 25-30.

———, The chromatic class of a multigraph (Russian), Kibernetika (Kiev) 3 (1965), 29-39.

———, Critical graphs with a given chromatic class (Russian), Diskret. Analiz 5 (1965), 9-17.

———, Some unsolved problems in graph theory, Unspekhi Mat. Nauk 23 (1968), 117-134 = Russian Math. Surveys 23 (1968), 125-142.

W. D. Wallis, On one-factorizations of complete graphs, J. Austral. Math. Soc., 16 (1973), 167-171.

M. E. Watkins, A theorem on Tait colorings with an application to the generalized Petersen graphs, J. Combin. Theory, Ser.B, 6 (1969), 152-164.

R. J. Wilson, Some conjectures on edge-colorings of graphs, Recent Advances in Graph Theory (Ed., M. Fiedler), Academia, Prague (1975), 525-528.

———, Edge-colorings of graphs : A survey, Theory and Applications of Graphs (Eds., Y. Alavi and D. R. Lick), Proc. Internat. Conf. Western Mich. Univ., Springer-Verlag Lecture Notes in Math., Vol. 642 (1978), 608-619.

H. P. Yap, On the Critical Graph Conjecture, J. Graph Theory 4 (1980), 309-314.

———, A construction of chromatic index critical graphs, J. Graph Theory 5 (1981), 159-163.

———, On graphs critical with respect to edge-colorings, Discrete Math., 37 (1981), 289-296.

3. SYMMETRIES OF GRAPHS

1. The automorphism group of a graph

Suppose G and H are two graphs. If $\phi : V(G) \to V(H)$ is an injection such that $xy \in E(G)$ implies that $\phi(x)\phi(y) \in E(H)$ for any edge xy of G, then ϕ is called a monomorphism from G to H. If there exists a monomorphism from G to H, then we say that G is embeddable in H. Two graphs G and H are isomorphic if there is a bijection $\phi : V(G) \to V(H)$ such that $xy \in E(G)$ if and only if $\phi(x)\phi(y) \in E(H)$. An isomorphism from G onto itself is called an automorphism of G. The set of all automorphisms of G forms a group under the composition of maps and is denoted by Aut G, A(G) or simply by A. Thus we can consider A(G) as a group acting on V(G) which preserves adjacency. The automorphism group A(G) of G measures the degree of symmetry of G. If A(G) is the identity group, then G is called an asymmetric graph.

It was known to Riddell [51] that for large integers p, almost all labelled graphs of order p are asymmetric. Erdös and Rényi [63] gave a proof of this result using probabilistic methods. An outline of a proof of this result can also be found in Harary and Palmer [73; p. 206]. Wright [71,74] proved the same result for unlabelled graphs and Bollobás [82] generalized Wright's result to unlabelled regular graphs.

Suppose G is a graph. If G is asymmetric, then any vertex of G is distinguishable from all its other vertices. The other extreme situation is that for any two vertices x and y of G, there is an automorphism ϕ of G such that $\phi(x) = y$, in other words, the automorphism group A(G) of G acts transitively on the vertex set V(G) of G. In this case, we say that G is vertex-transitive (or vertex-symmetric).

It was pointed by E. Artin (Geometric Algebra, Interscience, New York, 1957) that "the investigation of symmetries of a given mathematical structure has always yield the most powerful results". Graphs form a nice mathematical structure. Many results on vertex-transitive graphs have been shown to be very useful in the construction of finite simple groups and block designs etc. The mean objectives of this chapter are to study various general properties of vertex-transitive graphs, and to classify the set of all vertex-transitive graphs according to different kinds of actions of the automorphism group on the vertex set. The following is a summary of the results presented in this chapter.

In §1 (an introductory section), we prove a simple relationship between vertex-transitivity and edge-transitivity.

In §2, we study the degree of asymmetry of asymmetric graphs.

In §3, we prove that for any given group Γ, there exists a graph G such that $A(G) \simeq \Gamma$ and we also mention several generalizations of this result.

In §4, we prove that every Cayley graph is vertex-transitive and that every vertex-transitive graph has a group-coset graph representation. We also mention some results on the graphical regular representation of a group.

In §5, we prove that every vertex-transitive graph of prime order is a circulant graph and we also give a formula for enumerating all circulant graphs of a given prime order p.

In §6, we show that any graph G can be auto-extended to a finite Cayley graph which preserves the chromatic number.

In §7, an elegant proof (due to Weiss) of Tutte's famous theorem on s-transitive cubic graphs is given.

In §8, we give a partial characterization of 4-ultratransitive graphs.

In §9, we discuss some progress made towards the resolution of Lovász' question which asks whether or not every connected Cayley graph is Hamiltonian.

In §10, we briefly mention some other interesting and important results about symmetries of graphs which we have not been able to disucss in detail due to lack of space.

Suppose G is a graph. It is clear that each ϕ in A(G) induces a bijection on E(G). If A(G) acts transitively on E(G), we say that G is edge-transitive (edge-symmetric).

The graph given in Fig.3.1(a) is vertex-transitive but not edge-transitive. The graph given in Fig.3.1(b) is edge-transitive but not vertex-transitive.

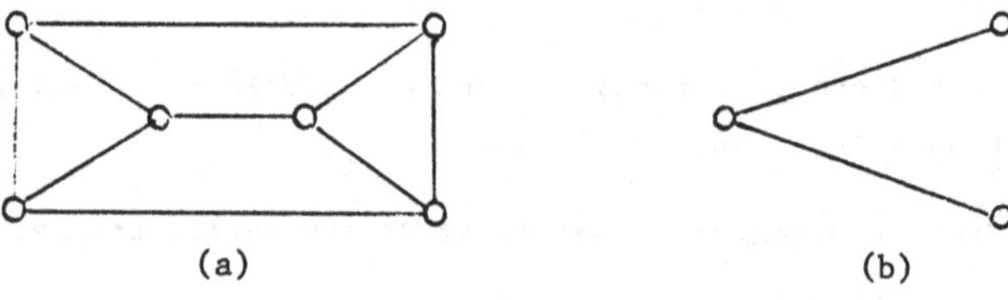

(a) (b)

Figure 3.1

Two vertices x and y of G are said to be similar if there exists an automorphism ϕ of G such that $\phi(x) = y$. It is clear that if two vertices x and y are similar, then $d(x) = d(y)$ and $G - x \cong G - y$. However, the converse is not true. An example can be found in Harary [69; p.171].

Theorem 1.1 below tells us a relationship between vertex-transitive graphs and edge-transitive graphs. This theorem, according to Harary [69;p.172], is due to Elayne Dauber. The proof given here is reproduced from Harary [69; p.172].

Theorem 1.1 If G is an edge-transitive graph with no isolated vertices, then either (i) G is vertex-transitive, or (ii) G is bipartite and G has two vertex-orbits which form the bipartition of G.

Proof. Let $e = uv \in E(G)$, and let V_1 and V_2 denote the orbits of u and v under the action of A(G) on V(G). Since G is edge-transitive, $V_1 \cup V_2 = V(G)$.

(i) If $V_1 \cap V_2 \neq \phi$, then G is vertex-transitive.

Let x and y be any two vertices of G. If $x, y \in V_1$, say, then there exist $\phi, \psi \in A(G)$ such that $\phi(u) = x$ and $\psi(u) = y$. Thus the automorphism $\psi\phi^{-1}$ is such that $\psi\phi^{-1}(x) = y$. If $x \in V_1$ and $y \in V_2$, let $w \in V_1 \cap V_2$. Since w is similar to both x and y, x and y are similar to each other.

(ii) If $V_1 \cap V_2 = \phi$, then G is bipartite.

Consider two vertices x and y in V_1. If they are adjacent, then there is $\psi \in A(G)$ such that $\psi(e) = xy$, which implies that one of the two vertices x and y is in V_1 and the other is in V_2, a contradiction. Hence G is bipartite. //

Corollary 1.2 If an edge-transitive graph G is regular of degree at least one and of odd order, then G is vertex-transitive.

Proof. If G is bipartite, then $|G|$ is even. //

Corollary 1.3 If G is edge-transitive and regular of degree $d \geq |G|/2$, then G is vertex-transitive.

From the above corollaries, we know that if G is edge-transitive, regular of degree d and not vertex-transitive, then $p = |G|$ is even, and $d < p/2$. Folkman [67] has proved that for each $p > 20$ divisible by 4, there exists a regular graph G of order p which is edge-transitive but not vertex-transitive.

Exercise 3.1

1. Show that for any graph G, $A(G) \cong A(\overline{G})$.
2. Show that the graph given in Fig.3.1(a) is vertex-transitive but not edge-transitive.
3. Suppose G is a graph of order p. Prove that $A(G) = \Sigma_p$, the symmetric group of degree p, if and only if $G = K_p$ or O_p.
4. Prove that $A(C_p) = D_p$, the dihedral group of degree p.
5. Find the automorphism group of the following graph

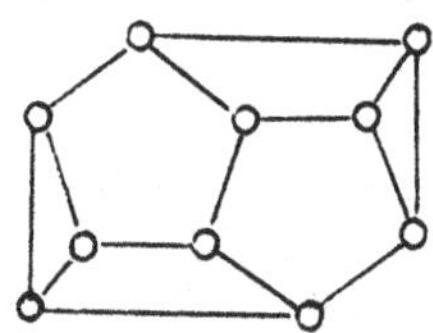

(Behzad and Chartrand [71;p.177]).

6. Prove that every connected, vertex-transitive graph of finite order has no cut-vertices.

7. A graph G is said to be symmetric if it is both vertex-symmetric and edge-symmetric. A graph G is said to be 1-transitive if for any two edges xy and uv of G, there is an automorphism ψ of G such that $\psi(x) = u$ and $\psi(y) = v$. Prove that every symmetric, connected graph of odd degree is 1-transitive (Tutte [66;p.59]).

(Note that Bouwer [70] proved that for any integer $n > 2$, there exists a connected graph which is symmetric but not 1-transitive. Holt [81] has constructed a graph of order 27, regular of degree 4, which is symmetric but not 1-transitive. This graph is the Cayley graph $G = G(\Gamma,S)$ where the group $\Gamma = \langle \alpha,\beta \mid \alpha^9 = \beta^3 = 1, \beta^{-1}\alpha\beta = \alpha^4 \rangle$ and $S = \{\beta\alpha, \beta\alpha^{-1}, \beta^2\alpha^2, \beta^2\alpha^{-2}\}$. For definition of a Cayley graph, see §4.)

8. Let G and H be two graphs. If $\phi : E(G) \to E(H)$ is a bijection such that if any two edges e and f are incident in G, $\phi(e)$ and $\phi(f)$ are also incident in H, then ϕ is called an edge-isomorphism from G to H.

Prove that if ϕ is an edge-isomorphism from a connected graph G to a connected graph H, where $G \neq K_3$, S_4, K_4 or one of the following two graphs, then ϕ is induced by an isomorphism from G to H (Whitney [32], see also Behzad, Chartrand and Lesniak-Foster [79; p.177]).

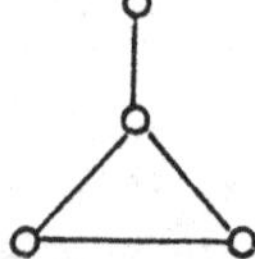
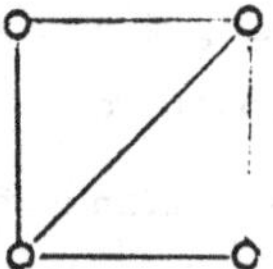

9†. Let G be an edge-transitive graph which is regular of degree d. Prove that if $|G| = 2p$ or $2p^2$, where p is a prime, then G is vertex-transitive (Folkman [67]).

2. Asymmetric graphs

In section 1 we have mentioned that for large p, almost all graphs of order p are asymmetric. Thus we shall first study some properties of asymmetric graphs before we study symmetries in graphs.

Evidently any asymmetric graph can be turned into a non-asymmetric graph by deleting some of its edges and by adding some new edges joining its vertices. We call such a transformation of a graph G a *symmetrization* of G. Suppose in a symmetrization of G, r edges have been deleted and s edges have been added. We define the *degree of asymmetry* a(G) of G as the minimum of r + s where the minimum is taken over all possible symmetrizations of G. Hence a graph G is asymmetric if and only if $a(G) > 0$.

For a fixed positive integer n, we denote by A(n) the maximum of a(G) for all graphs G of order n. By convention, we define $A(1) = +\infty$. It is obvious that $A(2) = A(3) = 0$. Using the fact that for any Graph G, $a(G) = a(\bar{G})$, we can prove that $A(4) = A(5) = 0$ (see Ex.3.2(1)).

We now prove the main theorem of this section.

Theorem 2.1 (Erdös and Rényi [63]) $A(n) \leq [\frac{n-1}{2}]$.

Proof. It is clear that for $n \leq 3$, $A(n) = 0$. Hence we suppose $n \geq 4$. Let G be a graph having n vertices $v_1, v_2, \ldots, v_n$ and let d_k be the valency of v_k. For two distinct vertices v_j and v_k, let f_{jk} be the number of vertices v_i which are adjacent to both v_j and v_k. We also define $f_{jj} = 0$. By definition, it is clear that $f_{jk} = f_{kj}$. Since $\sum_{j=1}^{n} \sum_{k=1}^{n} f_{jk}$ is the number of ordered pairs of edges of G which have one common vertex, we have

$$\sum_{j=1}^{n} \sum_{k=1}^{n} f_{jk} = \sum_{i=1}^{n} d_i(d_i - 1) \tag{1}$$

For $j \neq k$, we also define

$$\Delta_{jk} = d_j + d_k - 2f_{jk} - 2\delta_{jk} \tag{2}$$

where $\delta_{jk} = 1$ or 0 according to whether v_j and v_k are adjacent or not. We further define $\Delta_{jj} = 0$. Evidently, Δ_{jk} is the number of vertices which are adjacent to either v_j or v_k but not both.

Now by deleting all the Δ_{jk} edges joining to either v_j or v_k but not both, we obtain a graph G' in which v_j and v_k are similar. Thus

$$A(n) \le \min_{j \ne k} \Delta_{jk} \le \frac{1}{n(n-1)} \sum_{j=1}^{n} \sum_{k=1}^{n} \Delta_{jk} \tag{3}$$

Next, from (1) and (2) we have

$$\sum_{j=1}^{n} \sum_{k=1}^{n} \Delta_{jk} = 2 \sum_{i=1}^{n} d_i(n - 1 - d_i) \tag{4}$$

It is clear that

$$d_i(n - 1 - d_i) = \left(\frac{n-1}{2}\right)^2 - \left(d_i - \frac{n-1}{2}\right)^2 \tag{5}$$

From (3), (4) and (5), we obtain

$$A(n) \le \begin{cases} \dfrac{n-1}{2} & \text{if } n \text{ is odd} \\ \dfrac{n(n-2)}{2(n-1)} & \text{if } n \text{ is even} \end{cases} \tag{6}$$

However, since $n > 4$, $\frac{n(n-2)}{2(n-1)} < \frac{n}{2}$. Consequently, $A(n) \le [\frac{n-1}{2}]$. //

It is clear that if there exists a graph G of odd order n such that $a(G) = \frac{n-1}{2}$, then

$$\min_{j \ne k} \Delta_{jk} = \frac{n-1}{2} \tag{7}$$

As by (3), (4) and (5), we have for odd n,

$$\min_{j \ne k} \Delta_{jk} \le \frac{1}{n(n-1)} \sum_{j \ne k}\sum \Delta_{jk} \le \frac{n-1}{2} \ .$$

It follows that (7) can hold only if $\Delta_{jk} = \frac{n-1}{2}$ for all $j \ne k$. In this case, it follows from (5) that we also have $d_i = \frac{n-1}{2}$ for all $i = 1, 2, \ldots, n$. Now if $n \equiv 3 \pmod 4$, then $\frac{n-1}{2}$ is odd. This contradicts the fact that in any graph the number of vertices having odd valency is even. Thus (7) can hold for an odd n only if $n \equiv 1 \pmod 4$.

We shall call a graph G of order n, $n \equiv 1 \pmod 4$, for which (7) holds a <u>Δ-graph</u>. Erdös and Rényi [63] have constructed many Δ-graphs. Their graphs constructed in this connection are all both vertex-transitive and edge-transitive. This indicates that, although most graphs are asymmetric, the class of graphs having many similar vertices also plays an important role in graph theory. Further examples are Turan's graphs $T_r(n)$ (see Bondy and Murty [76;p.110]) which also have many similar vertices.

Conjecture (Erdös and Rényi [63]) <u>The automorphism groups of all Δ-graphs are nontrivial.</u>

Exercise 3.2

1⁻. Prove that $A(4) = A(5) = 0$.

2. Construct four asymmetric graphs of order 6 each having size at most 7. Determine the degree of asymmetry of each of these four graphs.

3. Let n and k be two positive integers such that $n > 2k + 1$. Let $C(n,k)$ denote the least value such that there exists a connected, asymmetric graph G of order n, having $C(n,k)$ edges, and $a(G) = k$. Prove that $C(n,1) = n - 1$ for $n \geq 7$.

4⁺. Let C^t be the class of connected graphs having no vertices of valency 2. Let $C^t(n,1)$ denote the least value such that there exists an asymmetric graph G in C^t having n vertices, $a(G) = 1$ and $C^t(n,1)$ edges, where the non-asymmetric graph obtained from G by applying a symmetrization to G is again in C^t. Prove that $C^t(7,1) = 11$, and

$$C^t(n,1) = \begin{cases} n + 2 & \text{for } n = 8 + 2p,\ p = 0, 1, 2, \ldots \\ n + 1 & \text{for } n = 9 + 2p,\ p = 0, 1, 2, \ldots \end{cases}$$

(Quintas [67; Theorem 9]).

5. Prove that $a(T) \leq 1$ for any tree T.

6⁺. Prove that, for large n, the automorphism groups of almost all trees of order n are nontrivial (Erdös and Rényi [63]).

7⁺. Let $F(n,k)$ be the smallest integer such that there exists an asymmetric graph G with n vertices, $a(G) = k$ and $F(n,k)$ edges. Prove that $C(n,2) > F(n,2)$ for $n \geq 21$ (Baron [70]).

(Note that Shelah [73] proved that (i) for sufficiently large n, $F(n,2) = n + 1$, $C(n,2) = n + 2$; for odd $k > 2$, and n sufficiently larger than k, $F(n,k) = C(n,k) = [(k + 3)n/4 - 0.5[2n/(k + 3)] + \frac{1}{2}]$; and (iii) for even $k > 2$ and n sufficiently larger than k, $F(n,k) = C(n,k) = [(k + 2)n/4 + \frac{1}{2})]$.)

8.$^{+}$ Let $m(n)$ ($M(n)$) be the minimum (maximum) degree of regular asymmetric graphs with n vertices. For $k > 0$, let $p(k)$ be the least number of vertices of asymmetric, regular graphs of degree k. Prove that (i) $m(n) = 4$, $M(n) = n - 5$ for odd $n > 10$ and $m(n) = 3$, $M(n) = n - 4$ for even $n > 10$; (ii) $p(k) = k + 4$ for even $k > 6$, and $p(k) = k + 5$ for odd $k > 6$ (Baron and Imrich [69]).

3. Graphs with a given group

The first book on graph theory was written by König in 1936. In this book [p.5], König proposed the problem of determining all finite groups Γ for which there exists a graph G such that $A(G) \cong \Gamma$. Two years later, this problem was solved by Frucht [38] who proved that every finite group has this property. In this section we shall present a proof of Frucht's theorem and mention a few further development of this problem. Much material of this section are taken from Behzad and Chartrand [71; p.173-179]. The survey papers by Babai [81] and Holton [76] are recommended for further information on this topic.

With every finite group $\Gamma = \{\phi_1,\ldots,\phi_p\}$, we can construct a complete symmetric digraph $D = D(\Gamma)$ (a <u>complete symmetric digraph</u> D is a digraph such that for any two vertices u and v of D, both (u,v) and (v,u) are arcs of D) having vertex set $V(D) = \{\phi_1,\ldots,\phi_p\}$. Each arc (ϕ_i,ϕ_j) of D is labelled $\phi_i^{-1}\phi_j$ (i.e. coloured with colour $\phi_i^{-1}\phi_j$). An element $\alpha \in A(D)$ is said to be <u>colour-preserving</u> if the arcs (ϕ_i,ϕ_j) and $(\alpha(\phi_i), \alpha(\phi_j))$ have the same colour for every arc (ϕ_i,ϕ_j) of D. It is clear that the set of colour-preserving automorphisms of D forms a subgroup of A(D).

Lemma 3.1 <u>Every finite group</u> Γ <u>is isomorphic to the group of colour-preserving automorphisms of</u> $D(\Gamma)$.

Proof. Let $\Gamma = \{\phi_1,\phi_2,\ldots,\phi_p\}$ where ϕ_1 is the identity. For $i = 1,2,\ldots,p$, define $\alpha_i : \Gamma \to \Gamma$ by $\alpha_i(\phi_m) = \phi_i\phi_m$ for all m. The label of an arc (ϕ_r,ϕ_s) is the same as that of $(\alpha_i(\phi_r),\alpha_i(\phi_s))$ since $\phi_r^{-1}\phi_s = (\phi_i\phi_r)^{-1}(\phi_i\phi_s)$. Hence each α_i is a colour-preserving automorphism of $D(\Gamma)$. Also, $\alpha_i \neq \alpha_j$ for $i \neq j$.

We now show that if α is a colour-preserving automorphism of $D(\Gamma)$, then $\alpha = \alpha_i$ for some $i = 1,2,\ldots,p$. Suppose $\alpha(\phi_1) = \phi_i$. For each $m \neq 1$, let $\alpha(\phi_m) = \phi_{j(m)}$. The label of (ϕ_1,ϕ_m) is ϕ_m, therefore the label of $(\alpha(\phi_1),\alpha(\phi_m))$ is ϕ_m also, i.e. $\phi_i^{-1}\phi_{j(m)} = \phi_m$. Hence $\alpha(\phi_m) = \phi_i\phi_m$ and $\alpha = \alpha_i$.

Finally, we prove that $\psi : \phi_i \to \alpha_i$ defines a group isomorphism. It is clear that ψ is an injection. It remains to show that ψ is a group homomorphism. Let $\phi_i\phi_j = \phi_k$. Then for each $\phi_m \in \Gamma$, $\alpha_i\alpha_j(\phi_m) = \alpha_i(\alpha_j\phi_m) = \alpha_i(\phi_j\phi_m) = \phi_i(\phi_j\phi_m) = (\phi_i\phi_j)\phi_m = \phi_k\phi_m = \alpha_k(\phi_m)$. Thus $\psi(\phi_i\phi_j) = \psi(\phi_i)\psi(\phi_j)$. //

We shall now apply the above lemma to construct a graph G such that A(G) is isomorphic to a given finite group $\Gamma = \{\phi_1,\phi_2,\ldots,\phi_p\}$ where ϕ_1 is the identity. For $|\Gamma| = 1$ or 2, we take $G = K_1$ or K_2. Hence we assume that $|\Gamma| > 3$. We first construct the complete symmetric digraph $D(\Gamma)$. Now by Lemma 3.1, the group of colour-preserving automorphisms of $D(\Gamma)$ is isomorphic to Γ. We next transform the digraph $D(\Gamma)$ into a graph G in the following way. Suppose (ϕ_i,ϕ_j) is an arc of $D(\Gamma)$ which is labelled $\phi_i^{-1}\phi_j = \phi_k$. We delete this arc and replace it by a path ϕ_i, u_{ij}, v_{ij}, ϕ_j. At the vertex u_{ij} we add a new path P_{ij} of length $2k - 4$, and at the vertex v_{ij} a new path Q_{ij} of length $2k - 3$. This operation is performed with every arc of $D(\Gamma)$. The addition of the paths P_{ij} and Q_{ij} in the formation of G is, in a sense, equivalent to the preservation of direction of arcs of $D(\Gamma)$. Since every colour-preserving automorphism of $D(\Gamma)$ gives rise to an automorphism of G, and conversely, we have the following theorem.

Theorem 3.2 (Frucht's theorem) <u>For any finite group</u> Γ, <u>we can construct a graph</u> G <u>such that</u> $A(G) \cong \Gamma$.

Remarks.

1. The number $|G|$ of vertices of the graph G given in the proof of Theorem 3.2 is excessive. For instance, if Γ is a group of order 3, then $|G|$ given in the proof of Theorem 3.2 is 33. However, it can be shown that 9 is best possible for the group of order 3 (see Ex.3.3(1)).

(For discussions on the minimum order of a graph G such that $A(G) \cong \Gamma$, see Sabidussi [59]. For results on upper and lower bounds of the order of G such that $A(G) \cong \Gamma$, see Babai [81].)

2. Many people have constructed other types of graphs G such that $A(G) \cong \Gamma$ where G satisfies some additional properties. Examples are as follows:

(i) For any finite group Γ, there exist infinitely many cubic graphs G such that $A(G) \cong \Gamma$ (Frucht [49]).

(As mentioned in Babai [81], there are some gaps in the original proof. for an alternate proof see Lovász [79; Chap.2, Problem 8]. This result has been further generalized by Sabidussi [57].)

(ii) Given a finite group Γ, and integers n, χ, c such that $3 \leq n \leq 5$, $2 \leq \chi \leq n$, $1 \leq c \leq n$, there exist infinitely many graphs G which are regular of degree n, have chromatic number χ, connectivity c, and are such that $A(G) \cong \Gamma$ (Izbicki [57,59]).

3. The result of Theorem 3.2 is sometimes called the abstract finite group version of König's question. Some authors also considered the permutation group version of König's question: "For which permutation group Γ does there exist a graph G such that $A(G) \equiv \Gamma$, where $\equiv$ means that A(G) and Γ are isomorphic as permutation groups?" It is obvious that not all permutation groups (for instance, the cyclic group of order 3) are isomorphic with A(G) as permutation groups, for some graph G. We mention some results here.

(i) Suppose $\Gamma = \Gamma_1 \times \Gamma_2$ (the direct product of Γ_1 and Γ_2) or $\Gamma = A_n$, where Γ_1 is the cyclic group generated by $(1,2,\ldots,n)$, Γ_2 is any permutation group on $\{0,1,2,\ldots,m\}$ and A_n, for $n \geq 2$, is the alternating group. Then there is no graph G with $A(G) \equiv \Gamma$ (Kagno [46,47,55]; see also Chao [64]).

(Some of the above results have been generalized by Alspach [74].)

(ii) Suppose Γ is the cyclic group generated by $(x_{11},x_{12},\ldots,x_{1m})\ldots(x_{n1},x_{n2},\ldots,x_{nm})$, $m > 5$, $n \geq 2$ or $(u_1)\ldots(u_m)(v_1,v_1')\ldots(v_n,v_n')$, $n \geq 1$. Then there exists a graph G such that $A(G) \equiv \Gamma$ (Mohanty, Sriharan and Shukla [78]).

4. Cameron [80b] proved if Γ is a finite group and $C(\Gamma)$ is the class of graphs G for which Γ is a subgroup of A(G), then there is a rational number $a(\Gamma)$ with $0 \leq a(\Gamma) \leq 1$ such that the proportion of graphs of order p in $C(\Gamma)$ which satisfy $\Gamma = A(G)$ tends to $a(\Gamma)$ as $p \to \infty$. Moreover,

(i) $a(\Gamma) = 1$ if and only if Γ is a direct product of symmetric groups;

(ii) if Γ is abelian, then $a(\Gamma) = 0$ or 1;

(iii) the values of $a(\Gamma)$ for metabelian groups Γ are dense in the interval [0,1].

Exercise 3.3

1. Let Γ be a group of order 3. Construct a graph G having 9 vertices such that $A(G) \cong \Gamma$. Show that 9 is best possible.

2. Construct a cubic graph G such that $|A(G)| = 1$.

3$^+$. Prove that for any finite group Γ and any integers n and χ where $n \geq 3$ and $2 \leq \chi \leq n$, there exist infinitely many graphs G such that $A(G) \cong \Gamma$, $\chi(G) = \chi$, and G is regular of degree n (Izbicki [60]).

4$^+$. Prove that for any infinite group Γ, there exists a connected graph G such that $A(G) \cong \Gamma$ (Sabidussi [60]).

5. Prove that if Γ is a group of order $n \geq 6$, then there exists a graph G having at most 2n vertices such that $\Gamma \cong A(G)$ (Babai [74]).

4. Vertex-transitive graphs

We first present a standard construction of VT-graphs (vertex-transitive graphs) due originally to Cayley [1878]. This construction is similar to, but more general than, the construction of the complete symmetric digraph $D(\Gamma)$ given in the previous section.

Suppose Γ is a group and $S \subseteq \Gamma$ is such that the identity $1 \notin S$ and $S^{-1} = \{x^{-1} \mid x \in S\} = S$. The Cayley graph $G = G(\Gamma,S)$ is the simple graph having vertex set $V(G) = \Gamma$ and edge set $E(G) = \{\{g,h\} \mid g^{-1}h \in S\}$. (In this section, in order to avoid confusion between the edge joining two vertices g and h with the group element gh, we shall write the edge joining two vertices g and h as $\{g,h\}$.) From this definition, it is obvious that a Cayley graph $G(\Gamma,S)$ is (i) complete if and only if $S = \Gamma^* = \Gamma - \{1\}$; (ii) connected if and only if S generates Γ. For more similar necessary and sufficient conditions of various properties of Cayley graphs $G(\Gamma,S)$ in terms of S, the readers may consult Teh and Shee [76].

Theorem 4.1 The Cayley graph $G = G(\Gamma,S)$ is vertex-transitive.

Proof. For each g in Γ we define a permutation ϕ_g of $V(G) = \Gamma$ by the rule $\phi_g(h) = gh$, $h \in \Gamma$. This permutation ϕ_g is an automorphism of G, for $\{h,k\} \in E(G) \Rightarrow h^{-1}k \in S \Rightarrow (gh)^{-1}(gk) \in S \Rightarrow \{\phi_g(h),\phi_g(k)\} \in E(G)$.

Now for any $h, k \in \Gamma$, $\phi_{kh^{-1}}(h) = (kh^{-1})h = k$. Hence G is vertex-transitive. //

The following is an example of a vertex-transitive graph which is not a Cayley graph.

Example The Petersen graph is vertex-transitive but it is not a Cayley graph.

Proof. It is clear that the diameter of a Cayley graph $G = G(\Gamma,S)$ is the smallest positive integer n such that

$$\Gamma = S \cup S^2 \cup \ldots \cup S^n$$

where $S^2 = \{hk \mid h, k \in S\}$ and $S^i = S^{i-1}S$ for $i \geq 3$. The diameter of the Petersen graph is 2. We now show that all the Cayley graphs of order 10 having degree 3 are of diameter greater than 2 and so none of them is the Petersen graph.

There are two groups of order 10. The first one is the cyclic group Z_{10} and the second one is the dihedral group D_5. The group operations here are additions and we replace S^{-1} by $-S$.

Case 1. $\Gamma = Z_{10} = \{0,1,\ldots,9\}$.

Since $-S = S$ and $|S| = 3$, $5 \in S$ and S can only be one of the following four sets

$$S_1 = \{1,5,9\}, \quad S_2 = \{2,5,8\}, \quad S_3 = \{3,5,7\}, \quad S_4 = \{4,5,6\}.$$

Now $|S_i + S_i| = 5$ for each $i = 1,2,3,4$. Thus the diameter of G is greater than 2.

Case 2. $\Gamma = D_5 = \{0,b,2b,3b,4b,a,a+b,a+2b,a+3b,a+4b\}$ where $2a = 0$, $5b = 0$ and $b + a = a + 4b$.

In this case a, $a + b$, $a + 2b$, $a + 3b$ and $a + 4b$ are the only elements of order 2 in Γ. Hence S can only be one of the following three types of sets

$$S_1 = \{a + jb, b, 4b\}, \quad j = 0,1,2,3,4\ ;$$

$$S_2 = \{a + jb, 2b, 3b\}, \quad j = 0,1,2,3,4\ ;$$

$$S_3 = \{a + j_1 b, a + j_2 b, a + j_3 b\}, \ 0 \leqq j_1 < j_2 < j_3 \leqq 4.$$

It can be verified that $|S_i + S_i| = 5$ for each $i = 1,2,3$. Thus the diameter of G is greater than 2 also. //

Although not every VT-graph is a Cayley graph, every VT-graph can be constructed almost like a Cayley graph. This result (Theorem 4.3) was due to Sabidussi [64]. We shall apply the following theorem to prove Theorem 4.3.

Theorem 4.2 Let H be a subgroup of a finite group Γ and let S be a subset of Γ such that $S^{-1} = S$ and $S \cap H = \phi$. If G is the graph having vertex set $V(G) = \Gamma/H$ (the set of all left cosets of H in Γ and edge set $E(G) = \{\{xH,yH\} \mid x^{-1}y \in HSH\}$, then G is vertex-transitive.

Proof. We first show that the graph G is well-defined. Suppose $\{xh,yH\} \in E(G)$ and $x_1H = xH$, $y_1H = yH$. Then $x_1 = xh$, $y_1 = yk$ for some $h, k \in H$. Now $x^{-1}y \in HSH \Rightarrow (xh)^{-1}(yk) \in HSH \Rightarrow x_1^{-1}y_1 \in HSH \Rightarrow \{x_1H,y_1H\} \in E(G)$. Hence the graph G is well-defined.

Next, for each $g \in \Gamma$ we define a permutation ψ_g of $V(G) = \Gamma/H$ by the rule : $\psi_g(xH) = gxH$, $xH \in \Gamma/H$. This permutation ψ_g is an automorphism of G, for $\{xh,yH\} \in E(G) \Rightarrow x^{-1}y \in HSH \Rightarrow (gx)^{-1}(gy) \in HSH \Rightarrow \{gxH,gyH\} \in E(G) \Rightarrow \{\psi_g(xH),\psi_g(yH)\} \in E(G)$.

Finally, for any xH, $yH \in \Gamma/H$, $\psi_{yx^{-1}}(xH) = yx^{-1}(xH) = yH$. Hence G is vertex-transitive. //

The graph G constructed in Theorem 4.2 is called the group-coset graph Γ/H generated by S and is denoted by $G(\Gamma/H,S)$.

Theorem 4.3 (Sabidussi's representation theorem) Let G be a vertex-transitive graph whose automorphism group is A. Let $H = A_b$ be the stabilizer of $b \in V(G)$. Then G is isomorphic with the group-coset graph $G(A/H,S)$ where S is the set of all automorphisms x of G such that $\{b,x(b)\} \in E(G)$.

Proof. It is easy to see that $S^{-1} = S$ and $S \cap H = \phi$. We now show that $\pi : A/H \to G$ given by $\pi(xH) = x(b)$, where $xH \in A/H$, defines a map. Suppose $xH = yH$. Then $y = xh$ for some $h \in H \Rightarrow \pi(yH) = y(b) = (xh)(b) = x(h(b)) = x(b) = \pi(xH)$.

We next show that π is a graph isomorphism :

π is one-to-one : Suppose $\pi(xH) = \pi(yH)$. Then $x(b) = y(b) \Rightarrow y^{-1}x(b) = b \Rightarrow y^{-1}x \in H \Rightarrow y \in xH \Rightarrow yH = xH$.

π is onto : Let c be a vertex of G. Since G is vertex-transitive, there exists z in A such that $z(b) = c$. Thus $\pi(zH) = z(b) = c$.

π preserves adjacency of vertices : $\{xH,yH\} \in E(G(A/H,S)) \Leftrightarrow x^{-1}y \in HSH \Leftrightarrow x^{-1}y = hzk$ for some $h, k \in H$, $z \in S \Leftrightarrow h^{-1}x^{-1}yk^{-1} = z \Leftrightarrow \{b,h^{-1}x^{-1}yk^{-1}(b)\} \in E(G) \Leftrightarrow \{b,x^{-1}y(b)\} \in E(G) \Leftrightarrow \{x(b),y(b)\} \in E(G) \Leftrightarrow \{\pi(xH),\pi(yH)\} \in E(G)$. //

Let A be the automorphism group of a vertex-transitive graph G. We know from a result mentioned in §2 of Chapter 1 that any two stabilizers A_b and A_c ($b, c \in V(G)$) are conjugate in A. Hence A_b is a normal subgroup of A if and only if $A_b = \{1\}$, where 1 is the identity of A. Thus the group-coset graph $G(A/H,S)$, where $H = A_b$, is a Cayley graph if and only if $A_b = \{1\}$, i.e. if and only if the action of A on V(G) is regular.

With the above motivation, we define the following term: Given a finite group Γ, if there exists a vertex-transitive graph G such that $A(G) \cong \Gamma$ and A(G) acts regularly on V(G), then G is called a graphical regular representation (GRR) of Γ, and Γ is said to have a graphical regular representation. We now prove the following theorem which is due to Chao [64] and Sabidussi [64].

Theorem 4.4 Let G be a vertex-transitive graph such that A(G) is abelian. Then A(G) acts regularly on V(G) and A(G) is an elementary abelian 2-group.

Proof. We first show that A(G) acts regularly on V(G). Suppose $g,h \in A(G)$, and g fixes $v \in V(G)$. Then $g(h(v)) = gh(v) = hg(v) = h(v)$ and so g fixes every vertex of G, because G is vertex-transitive. Hence $g = 1$, the identity of A(G).

Now, since A(G) acts regularly on V(G), G is isomorphic with the Cayley graph G(A(G),S). Also, as A(G) is abelian, $\psi : g \to g^{-1}$ is an automorphism of A(G), and it fixes S setwise. Hence, by Ex.4.4(2) and the fact that A(G) acts regularly on V(G), $g = \psi(g) = g^{-1}$ for every g in A(G) and so G is an elementary abelian 2-group. //

It has been proved that the group $(Z_2)^n$ has a GRR if and only if $n = 1$ or $n > 5$ (see McAndrew [65], Imrich [69,70] and Lim [78]). Hence, we have

Theorem 4.5 An abelian group Γ has a GRR if and only if $\Gamma = (Z_2)^n$ for $n = 1$ or $n > 5$.

It is clear that the cyclic group of order 3 has no GRR. The following interesting question arises naturally :

Which finite group Γ has a GRR?

Watkins [71] proved the following theorem.

Theorem 4.6 For any generating set S of a group Γ ($1 \notin S$, $S^{-1} = S$ and $\Gamma = \langle S \rangle$), if there exists a nontrivial group automorphism ϕ with $\phi(S) = S$, then Γ has no GRR.

This follows from Ex.3.4(2). Watkins conjectured that for finite groups Γ the converse of Theorem 4.6 is also true. Accordingly, he defines two classes of groups :

class I : Groups which have GRR's.

class II : Groups which have the property that any of its generating subsets is fixed by a nontrivial group automorphism.

As a consequence of Theorem 4.6, the two classes of groups are disjoint and Watkin's conjecture can be rephrased by saying that the union of class I and class II includes all finite groups.

A series of papers by Imrich, Nowitz and Watkins has been devoted to the resolution of the above conjecture. A complete answer to this conjecture is given by the following two theorems. These two theorems are deep and their proofs are difficult and long and therefore we refer the readers who are interested in this topic to their original papers.

Theorem 4.7 (Hetzel [76]) Every finite, non-abelian soluble group has a GRR except it is one of eight "exceptional groups".

Theorem 4.8 (Godsil [80]) Every finite non-soluble group has a GRR.

Combining Theorems 4.7 and 4.8, we have

Theorem 4.9 The union of class I and class II includes all finite groups.

Finally we give a representation theorem of symmetric graphs.

Theorem 4.10 (Teh and Chen [70])

(i) Let H be a subgroup of a finite group Γ, let $u \in \Gamma \setminus H$ and let $S = \{u, u^{-1}\}$. Then the group-coset graph $G = G(\Gamma/H, S)$ is symmetric.
(ii) Conversely, let G be a symmetric graph whose automorphism group is A, and let $H = A_b$ be the stabilizer of $b \in V(G)$. Then G is isomorphic with the group-coset graph $G(A/H, S)$ where $S = \{u, u^{-1}\}$ and $u \in A$ is such that $\{b, u(b)\} \in E(G)$.

Proof. (i) By Theorem 4.2, G is vertex-transitive and thus we need only to show that if $\{H,xH\}$, $\{H,yH\} \in E(G)$, then there exists an automorphism ψ of G such that $\psi\{H,xH\} = \{H,yH\}$.

Let $v = u$ or u^{-1}. Now $\{H,xH\},\{H,yH\} \in E(G) \Rightarrow x, y \in HSH$. We consider two cases.

Case 1. $x = avb$, $y = cvd$, where $a, b, c, d \in H$.

In this case, let $g = ca^{-1}$. Then $\psi_g(H) = gH = H$ and $\psi_g(xH) = gxH = yH$.

Case 2. $x = avb$, $y = cv^{-1}d$, where $a,b,c,d \in H$.

In this case, let $g = cv^{-1}a^{-1}$. Then $\psi_g(xH) = gxH = H$ and $\psi_g(H) = gH = yH$.

In either case, the automorphism ψ_g of G carries $\{H,xH\}$ to $\{H,yH\}$.

(ii) By the proof of Theorem 4.3, we know that $\pi : A/H \to V(G)$ given by $\pi(xH) = x(b)$ is a bijection and $\{xH,yH\} \in E(G(A/H,S)) \Rightarrow \{\pi(xH),\pi(yH)\} \in E(G)$. It remains to show that $\{\pi(xH),\pi(yH)\} \in E(G) \Rightarrow \{xH,yH\} \in E(G(A/H,S))$. Now $\{\pi(xH),\pi(yH)\} \in E(G) \Rightarrow \{x(b),y(b)\} \in E(G)$ $\Rightarrow \{b,x^{-1}y(b)\} \in E(G)$. Since G is edge-transitive, there exists $g \in A$ such that $g\{b,u(b)\} = \{b,x^{-1}y(b)\}$.

We consider two cases.

Case 1. $g(b) = b$ and $gu(b) = x^{-1}y(b)$.

In this case, $g, y^{-1}xgu \in H$. Hence $y^{-1}xgu = h$ for some $h \in H$ and so $x^{-1}y = guh^{-1} \in HSH$.

Case 2. $gu(b) = b$ and $g(b) = x^{-1}y(b)$.

In this case, $gu, y^{-1}xg \in H$. Hence $y^{-1}xg = guh$ for some $h \in H$ and so $y^{-1}x \in HSH$.

Hence, in either case, $\{xH,yH\} \in E(G(A/H,S))$. //

Exercise 3.4

1. Show that the set of all permutations ϕ_g of Γ in the proof of Theorem 4.1 forms a group Γ_1 which is isomorphic with Γ and is a subgroup of the group of automorphisms of $G(\Gamma,S)$.

2. Let $G = G(\Gamma,S)$ be a Cayley graph. Suppose π is an automorphism of the group Γ such that $\pi(S) = S$. Prove that π, regarded as a permutation of $V(G)$, is a graph automorphism fixing the vertex 1 where 1 is the identity of Γ. Show that the converse need not be true.

3. Prove that the group-coset graph $G(\Gamma/H,S)$ is complete if and only if $HSH = \Gamma \setminus H$.

4*. Is it true that every connected, vertex-transitive graph has a Hamilton path and every connected Cayley graph is Hamiltonian? (Lovász [70])

5. Let $G = G(\Gamma,S)$ be a Cayley graph and let H be a subgroup of Γ. Show that

   (i) if $H \cap S = \phi$, then $\chi(G) \leq |\Gamma|/|H|$; and

   (ii) if $H = \Gamma \setminus S$, then $\chi(G) = |\Gamma|/|H|$ (Teh [66]).

6. The <u>α-product (β-product)</u> of two graphs G and H is the graph $G \times_\alpha H$ ($G \times_\beta H$) with vertex set $V(G) \times V(H)$, in which (x,y), $(u,v) \in V(G) \times V(H)$ are adjacent if and only if $xu \in E(G)$ and $yv \in E(H)$ (either $x = u$ and $yv \in E(H)$ or $y = v$ and $xu \in E(G)$). Prove that the α-product and the β-product of two vertex-transitive graphs G and H are vertex-transitive (Teh and Yap [64]).

   (There are many methods by which we can construct other vertex-transitive graphs from a give collection of vertex-transitive graphs. For more details, see Teh and Yap [64].)

7^+. A digraph D is regular if the in-degree and the out-degree of all the vertices of D are equal.

   Let Q be a quasi-group and let $S \subseteq Q$. The <u>quasi-group digraph</u> $D = G(Q,S)$ is constructed as follows:

$$V(D) = Q, \quad E(D) = \{(x,xa) \mid x \in Q,\ a \in S\}.$$

   Prove that every finite quasi-group digraph is a regular digraph and every finite regular digraph is a quasi-group digraph (Teh [69]).

8. Let G be a vertex-transitive graph. Prove that if $A(G)$ contains a transitive abelian subgroup, then any two vertices in G are <u>involuntory</u>, i.e. for any two vertices u and v in G, there exists

$\sigma \in A(G)$ such that $\sigma(u) = v$ and $\sigma(v) = u$ (Bohdan [75]).

5. Vertex-transitive graphs of prime order

Cayley graphs $G(Z_n,S)$ are usually called circulant graphs. The main objectives of this section are :

(1) to prove that every VT-graph of prime order p is a circulant graph $G(Z_p,S)$;

(2) to find the automorphism group of the circulant graph $G(Z_p,S)$; and

(3) to enumerate the circulant graphs $G(Z_p,S)$.

Theorem 5.1 (Turner [67]) *If G is a vertex-transitive graph of prime order p, then it is a circulant graph.*

Proof. Let $V(G) = \{u_0, \ldots, u_{p-1}\}$, $A = A(G)$ and let H be the stabilizer of u_0. Suppose $\phi_i \in A$ is such that $\phi_i(u_0) = u_i$. Then $|A| = |H||Orb(u_0)| = p|H|$. Thus $p||A|$ and by Sylow's theorem, A contains a subgroup $K = \{1,\pi,\ldots,\pi^{p-1}\}$ of order p. Rename the vertices $\{u_0,\ldots,u_{p-1}\}$ as $\{v_0,\ldots,v_{p-1}\}$ so that $\pi(v_i) = v_{i+1}$, $0 \leq i \leq p-2$ and $\pi(v_{p-1}) = v_0$. Suppose $\{v_0,v_1\} \in E(G)$. Then $\{v_i,v_{2i}\} = \pi^i\{v_0,v_i\}$ (by definition, $\pi^i\{v_0,v_i\} = \{\pi^i(v_0), \pi^i(v_i)\}$), $\{v_{2i},v_{3i}\} = \pi^i\{v_i,v_{2i}\}$, $\ldots$, $\{v_{(p-1)i},v_0\} = \pi^i\{v_{(p-2)i},v_{(p-1)i}\} \in E(G)$. Thus $v_0v_iv_{2i}\ldots v_{(p-1)i}v_0$ forms a cycle in G. If we write v_i as i, and let $S = \{i \mid \{v_0,v_i\} \in E(G)\}$, then G is the circulant graph $G(Z_p,S)$. //

We now use Burnside's theorem on transitive permutation groups of prime degree p to determine the automorphism group of VT-graphs of prime order p.

Let p be an odd prime, let Z_p be the field of integers modulo p, and let Z_p^* be the multiplication group of nonzero elements of Z_p. Since Z_p^* is cyclic of order p - 1, it has, for each divisor n of p - 1, a unique subgroup of order n. A proof of the following theorem can be found in Passman [68;p.53].

Theorem 5.2 (Burnside) *Let Γ be a transitive permutation group of prime degree p on a set B. Then either Γ is doubly transitive or B can* be identified with the field Z_p in such a way that

$$\Gamma \subseteq \{T_{a,b} \mid a \in Z_p^*,\ b \in Z_p\} = T$$

where $T_{a,b}$ is the permutation of Z_p which maps x to $ax + b$.

(Note that T forms a group under the operation $T_{a,b} \cdot T_{c,d} = T_{ac,b+ad}$.)

The following theorem was implicit in the work of Sabidussi [64]. The proof given here is due to Alspach [73].

Theorem 5.3 Let $G = G(Z_p,S)$ be a VT-graph of prime order p. If $S = \phi$ or $S = Z_p^*$, then $A(G) = \Sigma_p$, the symmetric group of degree p, otherwise

$$A(G) = \{T_{a,b} \mid a \in H,\ b \in Z_p^*\}$$

where $H = H(S)$ is the largest even order subgroup of Z_p^* such that S is a union of cosets of H.

Proof. Suppose $A(G)$ is doubly transitive on $V(G)$. Then it is not difficult to show that $S = \phi$ or $S = Z_p^*$. If $S = \phi$ or $S = Z_p^*$, it is clear that $A(G) = \Sigma_p$.

Suppose $A(G)$ is not doubly transitive on $V(G)$. Then by Burnside's theorem,

$$A(G) \subseteq \{T_{a,b} \mid a \in Z_p^*,\ b \in Z_p\} = T.$$

It is clear that

$$J = \{a \mid T_{a,o} \in A(G)_o\}$$

where $A(G)_0$ is the stabilizer of 0, is a subgroup of even order of Z_p^* (Note that $T_{a,o} \in A(G)_o \Rightarrow T_{-a,o} \in A(G)_o$.) such that $JS = \{js \mid j \in J, s \in S\} = S$. Hence $A(G)_o \subseteq \{T_{a,o} \mid a \in H\}$. In fact, since each $T_{a,o}$, $a \in H$, maps adjacent vertices to adjacent vertices, we have

$$A(G)_o = \{T_{a,o} \mid a \in H\}.$$

Finally, since G is vertex-transitive,

$$A(G) = A(G)_o \cup T_{1,1} A(G)_o \cup \cdots \cup T_{1,p-1} A(G)_o$$
$$= \{T_{a,b} \mid a \in H,\ b \in Z_p\}. \qquad //$$

Corollary 5.4 If $G = G(Z_p,S)$ and $(p-1,|S|) = 2$, then $A(G) = D_p$, the dihedral group of degree p.

Proof. Since H is a subgroup of Z_p^* and S is a union of cosets of H, $|H|$ must divide $(p-1,|S|)$, the g.c.d. of $p - 1$ and $|S|$. Hence, if $(p-1,|S|) = 2$, then $|H| = 2$ and $|A(G)| = 2p$. Thus $A(G) = D_p$. //

A graph G is said to be symmetric if G is both vertex-symmetric and edge-symmetric. The following result was conjectured by Turner [67] and proved by Chao [71]. The proof given below is due to Berggren [72].

Theorem 5.5 Let p be an odd prime. The graph G with p vertices, each having valency $n > 2$, is symmetric if and only if $G = G(Z_p,H)$ where H is the unique subgroup of Z_p^* of order n.

Proof. Suppose $G = G(Z_p,H)$ where H is the unique subgroup of Z_p^* of even order n. Then for any $\{x,y\}$, $\{u,v\} \in E(G)$, there is $h \in H$ such that $(y - x)h = v - u$. Let $b \in Z_p$ be such that $yh + b = v$. Then $T_{h,b}$ is an automorphism of G such that $T_{h,b}\{x,y\} = \{u,v\}$.

Conversely, suppose G is symmetric. Then $G = G(Z_p,S)$ for some $S \subseteq Z_p$. By the proof of Theorem 5.3, $S = Hs_1 \cup \dots \cup Hs_r$ where H is the largest even order subgroup of Z_p^* such that S is a union of cosets of H.

If $r > 2$, then there is $T_{h,b} \in A(G)$, $h \in H$, that maps the edge $\{0,s_1\}$ to the edge $\{0,s_2\}$. In case $T_{h,b}(0) = 0$, we have $b = 0$ and $hs_1 = s_2$, from which it follows that $Hs_1 = Hs_2$, a contradiction. In case $T_{h,b}(0) = s_2$ and $T_{h,b}(s_1) = 0$, we have $(-h)s_1 = s_2$, from which it follows that $Hs_1 = Hs_2$ again. Hence $S = Hs_1$ for some $s_1 \in S$.

Finally, it can be shown that $\pi : x \to xs_1^{-1}$ is an isomorphism from $G(Z_p,S)$ to $G(Z_p,H)$. //

Remark. By Corollary 1.2, it would be enough to assume that in Theorem 5.5, G is regular and edge-symmetric.

Suppose $G = G(Z_p,S)$ is a circulant graph. The set S is called the symbol of G. Two symbols S and S' are equivalent if there exists a positve integer $q < (p-1)/2$ such that $S' = qS = \{qs \mid s \in S\}$. Since

$i \in S$ implies that $p - i \in S$, in practice, we need only to write out the elements i of S such that $i \leq (p-1)/2$.

Theorem 5.6 (Turner [67]) Suppose p is a prime. Two circulant graphs $G = G(Z_p, S)$ and $G' = G(Z_p, S')$ are isomorphic if and only if their corresponding symbols S and S' are equivalent.

Proof. Suppose $\pi : G' \to G$ is an isomorphism. Then $A(G') = \pi^{-1}A(G)\pi$.

By Theorem 5.3 and Sylow's theorem, both $A(G')$ and $A(G)$ have a unique subgroup K of order p. Hence $K = \pi^{-1}K\pi$. Now by multiplying π by an element in K (if necessary), we may assume that π fixes the vertex 0. Thus π carries S to S'. However, by Theorem 5.3, any automorphism of G fixing the vertex 0 has the form $z \to qz$ for some $q \in Z_p^*$. Since $qS = (p - q)S$, we may further choose q so that $1 \leq q \leq \frac{p-1}{2}$ and $qS = S'$. //

Theorem 5.6 shows that to count the number of non-isomorphic VT-graphs of prime order p, it suffices to count the number of non-equivalent symbols. The cycle index of the cyclic group C_p is given by

$$Z(C_p;\ x_1,\ x_2,\ \ldots,\ x_m) = \frac{1}{m} \sum_{d|m} \phi(d) x_d^{m/d}$$

where $m = (p-1)/2$ and $\phi(d)$ is the Euler ϕ-function. Now Polya's Theorem on Pattern Counting says that the enumerating polynomial for the circulant graphs of prime order p is given by

$$Z(C_p;\ 1 + x^p,\ \ldots,\ 1 + x^{mp})$$

where the coefficient of x^i is the number of non-isomorphic copies of size i.

Table of number of non-isomorphic VT-graphs of prime order p.

Prime	Enumerating polynomial	Total number of graphs
3	$1+x^3$	2
5	$1+x^5+x^{10}$	3

7	$1+x^7+x^{14}+x^{21}$	4
11	$1+x^{11}+2x^{22}+2x^{33}+x^{44}+x^{55}$	8
13	$1+x^{13}+3x^{26}+4x^{39}+3x^{52}+x^{65}+x^{78}$	14
17	$1+x^{17}+4x^{34}+7x^{51}+10x^{68}+7x^{85}+4x^{102}+x^{119}+x^{136}$	36
19	$1+x^{19}+4x^{38}+10x^{57}+14x^{76}+14x^{95}+10x^{114}+4x^{133}+x^{152}+x^{171}$	60

Exercise 3.5

1⁻. Write out all the non-equivalent symbols of $G(Z_{13},S)$ where $|S| = 6$.

2. Let G be a connected, edge-transitive graph having prime size q. Prove that (i) if G is regular, then $G = C_q$; (ii) if G is not regular, then G is a star S_{q+1} (Turner [67]).

3. If $S = \{1,2,3,5,8,10,11,12\}$ and $G = G(Z_{13},S)$, find A(G).

4. Let $G = G(Z_p,S)$ be a VT-digraph of prime order p with symbol S (-S need not be equal to S). Prove that if $S \neq \phi$ or Z_p^*, then

$$A(G) = \{T_{a,b} \mid a \in D(S),\ b \in Z_p\}$$

where D(S) is the largest subgroup of Z_p^* such that S is a union of cosets of D(S) (Note that D(S) need not be of even order here.) (Alspach [73]; Chao and Wells [73]).

6. Auto-extensions

Vertex-transitive graphs, when drawn with a high degree of space symmetry, look very beautiful. Suppose we now have a graph G which is, in general, not vertex-transitive. Can we extend G to a (finite) vertex-transitive graph H so that G is a <u>section graph</u> (an induced subgraph) of G? From an artist's point of view, this is an interesting problem. The answer to this problem is in the affirmative and the main objective of this section is to introduce several methods for constructing them.

We say that a graph H is an <u>auto-extension</u> of a graph G if H can be decomposed into a collection of section graphs G_1, G_2, ... such that each G_i is isomorphic with G, for any $i \neq j$, G_i and G_j have at most one vertex in common, and each edge of H lies in one G_i.

Example Each of the graphs in the second row of Fig.3.2 (note that the size of these graphs have been reduced considerably) is an (infinite) auto-extension of the graph given above it.

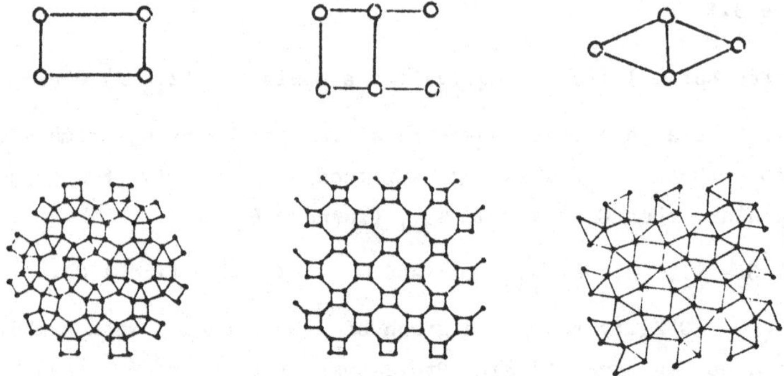

Figure 3.2

Theorem 6.1 (Ang and Teh [67]) *Every finite graph G can be auto-extended to a finite Cayley graph which preserves the chromatic number.*

Proof. Let $V(G) = \{v_0, v_1, \ldots, v_n\}$, $\chi(G) = m$ and let Z_{2m} be the group of integers under addition modulo 2m. It is clear that

$$H = \{(a_1, a_2, \ldots, a_n) \mid a_i \in Z_{2m}, \ i = 1, \ldots, n\}$$

(the direct sum of n copies of Z_{2m}) is an abelian group under component-wise addition. We shall now choose a suitable set $S \subseteq H$ such $H = G(H,S)$ is an auto-extension of G.

We first identify $v_0, v_1, \ldots, v_n$ with $(0,0,\ldots,0)$, $(1,0,\ldots,0)$, $(0,1,0,\ldots,0), \ldots, (0,0,\ldots,0,1)$ respectively. The set S is now defined as the set of all elements $s \in H$ such that $s = v_i - v_j$ where $v_iv_j \in E(G)$. From this construction, it is clear that for each $h \in H$, $G + h$ is a section graph of the Cayley graph H and $G + h \cong G$.

We now show that for any two distinct elements a, b in H, $|(G + a) \cap (G + b)| \leq 1$. To see this, we first note that $|(G + a) \cap$

$(G + b)| = |G \cap (G + b - a)|$. Thus, suppose $G \cap (G + c) \neq \phi$ for some $c = (c_1,\ldots,c_n) \in H$ with $c_i \neq 0$ and $x = (x_1,\ldots,x_n) \in G \cap (G + c)$. If $x = v_o$, then $c_i = -1$ for this particular i, and $c_j = 0$ for all $j \neq i$. Thus, for every $j \neq i$, $v_j + c \notin G$ and $|G \cap (G + c)| = 1$. On the other hand, if $x = v_i$, $i > 0$, then $v_i = v_j + c$ for some $v_j \in G$. Thus $c_i = 1$, $c_j = -1$ and $c_k = 0$ for all $k \neq i,j$. In this case, $v_k + c \notin G$ for any $k \neq j$ and thus $|G \cap (G +c)| = 1$ also.

Next, suppose $xy \in E(H)$. Then $x - y = v_i - v_j$ for some v_i, $v_j \in G$ such that $v_iv_j \in G$. Thus $x - v_i = y - v_j = h \Rightarrow x, y \in G + h$. Hence H is an auto-extension of G.

Finally, we prove that $\chi(H) = m = \chi(G)$.

Let ϕ be an m-vertex-colouring of G such that $\phi(v_o) = 0$ and $\phi(v_i) \in \{0,\ldots,m-1\}$. We shall extend ϕ to an m-vertex-colouring ψ of H. For this purpose, we consider H as a module over the ring $Z_m = \{0,1,\ldots,m-1\}$ and for each $a = (a_1,\ldots,a_n) \in H$, we define

$$\psi(a) = a_1\phi(v_1) + \ldots + a_n\phi(v_n)$$

where addition is taken modulo m.

We claim that for any two adjacent vertices a and b in H, $\psi(a) \neq \psi(b)$. Suppose otherwise. Let $a = (a_1,\ldots,a_n)$, $b = (b_1,\ldots,b_n)$. Then $\sum a_i\phi(v_i) = \sum b_j\phi(v_j)$. Now $ab \in E(H) \Rightarrow a - b = v_i - v_j$ where $v_iv_j \in E(G)$. Hence $a = b + v_i - v_j$, from which it follows that $a_i = b_i + 1$, $a_j = b_j - 1$, and $a_k = b_k$ for all $k \neq i$, j. Consequently, we have $(b_i + 1)\phi(v_i) + (b_j - 1)\phi(v_j) = b_i\phi(v_i) + b_j\phi(v_j) \Rightarrow \phi(v_i) = \phi(v_j)$, contradicting the assumption that $v_iv_j \in E(G)$. //

In general, $|H| = (2m)^n$ is too big. For instance, if $G = P_3$ is a path on 3 vertices, then $|H| = 16$. However, it is clear that C_6 (a cycle of order 6) is also an auto-extension of P_3 which preserves the chromatic number. Thus it is interesting to look for other extensions H such that the order of H is smaller. (Note that in the construction of H given in Theorem 6.1, if $m \geq 3$, Z_{2m} can be replaced by Z_m so that $|H|$ is considerably smaller.) An approach in this direction is given in Theorem 6.3 below.

We need the following definition and lemma. Let G be an additive group. A non-empty subset S of G is said to be parallelogram-free if for every $s_1, s_2, s_3, s_4 \in S$, $s_1 - s_2 = s_3 - s_4 \neq 0 \Rightarrow s_1 = s_3$ and $s_2 = s_4$.

Lemma 6.2 *If S is a parallelogram-free subset of a group G, then for any distinct elements x, y in G,* $|(x + S) \cap (y + S)| \leq 1$.

Proof. Suppose $|(x + S) \cap (y + S)| \geq 2$. Let $u, v \in (x + S) \cap (y + S)$, $u \neq v$. Then $u = x + s_1 = y + s_2$, $v = x + s_3 = y + s_4$ for some $s_1, s_2, s_3, s_4 \in S$. Hence $-x + y = s_1 - s_2 = s_3 - s_4 \neq 0$. Since S is parallelogram-free, $s_1 = s_3$ and $s_2 = s_4$, from which it follows that $u = v$, a contradiction. //

Theorem 6.3 (Tan and Teh [69]) *Every finite graph G can be auto-extended to a circulant graph of prime order.*

Proof. Let $V(G) = \{v_1, \ldots, v_k\}$ and let p be the smallest prime such that $p > 2^k + 1$. It is clear that

$$B = \{0, 1, 3, 7, \ldots, 2^i - 1, \ldots, 2^{k-1} - 1\}$$

is a parallelogram-free subset of Z_p.

Now for each $x \in Z_p$, we let $B_x = x + B$ and we turn B_x into a graph isomorphic to G by defining $V(B_x) = B_x$, and

$$\{x + 2^{i-1} - 1, x + 2^{j-1} - 1\} \in E(B_x) \text{ if and only if } v_i v_j \in E(G).$$

Next let H be the graph such that

$$V(H) = Z_p, \quad E(H) = \bigcup_{x \in Z_p} E(B_x).$$

Then since B is parallelogram-free, for any two distinct elements x and y of Z_p, $|B_x \cap B_y| \leq 1$. Also by the definition of E(H) we know that each edge of H lies in one B_x. Hence H is an auto-extension of G.

Finally we prove that H is a circulant graph. Suppose u and v are adjacent vertices of H. Then $u = x + 2^{i-1} - 1$ and $v = x + 2^{j-1} - 1$ for some $x \in Z_p$ and some indices i and j such that $v_i v_j \in E(G)$. Hence $\{x + 2^{i-1} - 1, x + 2^{j-1} - 1\} \in E(B_x) \Rightarrow \{w + x + 2^{i-1} - 1,$

$w + x + 2^{j-1} - 1\} \in E(B_{w+x})$ for any $w \in Z_p \Rightarrow$ the vertices $w + u$ and $w + v$ are adjacent in H. Thus H is a circulant graph. //

Remarks.

1. There are several generalizations of the constructions given by Theorems 6.1 and 6.3, see for instance, Chen and Teh [79].

2. Shee and Teh [84] have constructed other types of extensions of a finite graph to a finite vertex-transitive graph.

Exercise 3.6

1. Using the construction given in the proof of Theorem 6.1, find an auto-extension of P_3.

2. Prove that the converse of Lemma 6.2 is also true, i.e. if S is a subset of a group G such that for any two distinct elements x, y in G, $|(x + S) \cap (y + S)| \leq 1$, then S is a parallelogram-free subset of G.

3. Let J be the graph obtained from C_4 by joining two opposite vertices. Determine S, where S is the symbol of the circulant-graph $G(Z_{17},S)$ which is an auto-extension of J constructed by the method given in the proof of Theorem 6.3.

4. Prove that the graphs H constructed in Theorems 6.1 and 6.3 are regular of degree 2e(G).

7. s-transitive cubic graphs

In this and the following sections we shall restrict ourselves to a study of a few subclasses of the class of vertex-transitive graphs.

An <u>s-path</u> in a graph G is an (s+1)-tuple $[u] = (u_0, u_1, \ldots, u_s)$ of vertices of G with the property that $u_i u_{i+1} \in E(G)$ for all $i = 0, \ldots, s - 1$ and $u_i \neq u_{i+2}$ for $0 \leq i \leq s - 2$. A graph is <u>s-transitive</u> $(s \geq 0)$ if its automorphism group A(G) is transitive on the set of all s-paths in G and is <u>strictly s-transitive</u> if it is s-transitive but not (s+1)-transitive. For example, the complete bipartite graph of degree

$d > 2$ is 3-transitive but not 4-transitive. If G is s-transitive, then it is t-transitive for all $1 \leq t \leq s$. From the definition, a VT-graph is 0-transitive and a symmetric graph is 1-transitive. It is clear that s-transitive graphs are vertex-transitive and that the converse is not true in general.

Since every s-transitive graph is vertex-transitive, s-transitive graphs are regular graphs. The only connected VT-graphs of degree ≤ 2 are trivial s-transitive graphs. Hence we shall only confine ourselves to the study of nontrivial s-transitive graphs.

The main theorem of this section is

Theorem 7.1 (Tutte [47]) <u>Let</u> G <u>be a finite cubic graph</u>. <u>If</u> G <u>is s-transitive</u>, <u>then</u> $s \leq 5$.

The proof of this theorem is due to Tutte [47], with later improvements by Sims [67] and Djoković [72]. The following elegant proof of this theorem is due to Weiss [73]. Before we prove this theorem, let us first define some terminology.

Throughout this section, $G = (V,E)$ is a finite, connected, cubic graph which is strictly s-transitive. Let $\Gamma = A(G)$ and let $(x_0,\ldots,x_r)$ be an r-path in G. We set

$$\Gamma(x_0,x_1,\ldots,x_r) = \Gamma(x_0) \cap \ldots \cap \Gamma(x_r)$$

where $\Gamma(x_i)$ is the stabilizer of x_i. We shall write Π^X if Π is a group acting on the set X. If Σ is a subgroup of Γ, then the centre of Σ is denoted by $Z(\Sigma)$. We shall also write $\Sigma = 1$ if $\Sigma = \{1\}$ where 1 is the identity.

Lemma 7.2 $\Gamma(x_0,\ldots,x_s) = 1$ <u>for every</u> s-<u>path</u> $(x_0,\ldots,x_s)$ <u>in</u> G.

Proof. Let $[x] = (x_0,\ldots,x_s,x_{s+1})$ be an arbitrary (s+1)-path and let $X = N(x_s) - \{x_{s-1}\}$. Suppose $\Gamma(x_0,\ldots,x_s)^X$ is nontrivial. Since $|X| = 2$, $\Gamma(x_0,\ldots,x_s)$ is transitive on X. Let $[y] = (y_0,\ldots,y_s,y_{s+1})$ be another (s+1)-path . Since G is s-transitive, there exists $a \in \Gamma$ mapping $(y_0,\ldots,y_s)$ to $(x_0,\ldots,x_s)$. Now, as $a(y_{s+1}) \in X$, we may suppose that $a(y_{s+1}) = x_{s+1}$ by multiplying a with an appropriate element in

$\Gamma(x_0,\ldots,x_s)$, if necessary. Thus Γ contains elements mapping [y] to another (s+1)-path [x], which contradicts the assumption that G is not (s+1)-transitive. We conclude that $\Gamma(x_0,\ldots,x_s)^X = 1$ and $\Gamma(x_0,\ldots,x_s) = \Gamma(x_1,\ldots,x_{s+1})$.

Now let z be an arbitrary vertex of G. Since G is connected, there exists a path $(u_0,\ldots,u_t)$ with $u_0 = x_s$ and $u_t = z$. If $u_1 \neq x_{s-1}$, then $\Gamma(x_0,\ldots,x_s) = \Gamma(x_1,\ldots,x_s,u_1)$ as we have just shown. But then $\Gamma(x_1,\ldots,x_s,u_1) = \Gamma(x_2,\ldots,x_s,u_1,u_2)$ and so on. Hence $\Gamma(x_0,\ldots,x_s) \subseteq \Gamma(z)$. On the other hand, if $u_1 = x_{s-1}$, we extend [x] to an arbitrary (2s-1)-path $(x_0,\ldots,x_{s+1},x_{s+2},\ldots,x_{2s-1})$. Then

$$\begin{aligned}\Gamma(x_0,\ldots,x_s) &= \Gamma(x_1,\ldots,x_{s+1}) = \ldots = \Gamma(x_{s-1},\ldots,x_{2s-1})\\ &= \Gamma(u_1,u_0,x_{s+1},\ldots,x_{2s-1}) = \Gamma(u_2,u_1,u_0,x_{s+1},\ldots,x_{2s-2})\end{aligned}$$

and so on; again $\Gamma(x_0,\ldots,x_s) \subseteq \Gamma(z)$. Since z is arbitrary, we have $\Gamma(x_0,\ldots,x_s) = 1$. //

Lemma 7.3 $|\Gamma(x,y)| = 2^{s-1}$ for every 1-path (x,y).

Proof. Let $(x_0,x_1,\ldots,x_s)$ be an s-path in G with $x_0 = x$ and $x_1 = y$. Since $\Gamma(x_0,\ldots,x_s) = 1$, Γ acts regularly on the set W of all s-paths in G. Now $\Gamma(x_0,\ldots,x_s) \subseteq \Gamma(x_0,\ldots,x_{s-1})$ and $|N(x_{s-1}) - \{x_{s-2}\}| = 2$ imply $|\Gamma(x_0,\ldots,x_{s-1})| = 2$. Assume that $|\Gamma(x_0,\ldots,x_{s-i})| = 2^i$ for $1 < i < s - 2$. Then from $\Gamma(x_0,\ldots,x_{s-i}) \subseteq \Gamma(x_0,\ldots,x_{s-i-1})$, $|N(x_{s-i-1}) - \{x_{s-i-2}\}| = 2$, and the fact that Γ acts regularly on W, we have

$$|\Gamma(x_0,\ldots,x_{s-i-1})|/|\Gamma(x_0,\ldots,x_{s-i})| = 2.$$

The result follows by induction. //

In the following, we use $\partial(x,y)$ to denote the distance between two vertices x and y.

Lemma 7.4 Suppose $s > 2$. Let $x_0x_1 \in E$ and $y \in V$. If $\partial(x_i,y) < \frac{s}{2} - 1$ for $i = 0$ or 1, then

$$Z(\Gamma(x_0,x_1)) \subseteq \Gamma(y).$$

Proof. Let $(u_0,u_1,\ldots,u_s)$ be any s-path with either $u_0 = x_0$ and $u_1 = x_1$ or $u_0 = x_1$ and $u_1 = x_0$. It suffices to show that $Z(\Gamma(x_0,x_1)) \subseteq \Gamma(u_i)$ for all $i \leq s/2$. By Lemma 7.3, $\Gamma(x_0,x_1)$ is a nontrivial 2-group. Hence $Z(\Gamma(x_0,x_1)) \neq 1$. Since $\Gamma(u_0,\ldots,u_s) = 1$, there exists $t < s$ such that $Z(\Gamma(x_0,x_1)) \subseteq \Gamma(u_0,\ldots,u_t)$ but $Z(\Gamma(x_0,x_1)) \not\subseteq \Gamma(u_0,\ldots,u_{t+1})$. Let a be an element in $Z(\Gamma(x_0,x_1))$ but not in $\Gamma(u_0,\ldots,u_{t+1})$. We have

$$\Gamma(u_0,\ldots,u_{s-1}) = a\Gamma(u_0,\ldots,u_{s-1})a^{-1} = \Gamma(a(u_0),\ldots,a(u_{s-1}))$$
$$= \Gamma(u_0,\ldots,u_t,a(u_{t+1}),\ldots,a(u_{s-1}))$$

and so

$$\Gamma(u_0,\ldots,u_{s-1}) \subseteq \Gamma(u_{s-1},\ldots,u_t,a(u_{t+1}),\ldots,a(u_{s-1})).$$

Since $\Gamma(u_0,\ldots,u_{s-1}) \neq 1$ and $(u_{s-1},\ldots,u_t,a(u_{t+1}),\ldots,a(u_{s-1}))$ is a $2(s - t - 1)$-path, by Lemma 7.2, $2(s - t - 1) \leq s - 1$ and thus $s - 1 \leq 2t$. //

Lemma 7.5 _If_ s _is even, then_ $s \leq 4$.

Proof. Suppose $s > 4$. Let $m = \frac{s}{2} - 1$ and let $(x_0,\ldots,x_{3m})$ be an arbitrary 3m-path in G. Let $\Sigma_1 = Z(\Gamma(x_{m-1},x_m))$, $\Sigma_2 = Z(\Gamma(x_{2m},x_{2m+1}))$ and $\Sigma = [\Sigma_1,\Sigma_2] = \langle aba^{-1}b^{-1} \mid a \in \Sigma_1,\ b \in \Sigma_2\rangle$. By Lemma 7.4, $\Sigma_i \subseteq \Gamma(x_m,\ldots,x_{2m})$ for $i = 1,2$. Hence $\Sigma \subseteq \Gamma(x_m,\ldots,x_{2m})$. Suppose $2m \leq i \leq 3m$, $a \in \Sigma_1$ and $b \in \Sigma_2$. Then $\partial(x_{2m},a^{-1}(x_i)) = \partial(x_{2m},x_i) \leq m$ and so $b \in \Gamma(a^{-1}(x_i)) \cap \Gamma(x_i)$. Thus $aba^{-1}b^{-1}(x_i) = aba^{-1}(b^{-1}(x_i)) = ab(a^{-1}(x_i)) = a(a^{-1}(x_i)) = x_i$ and $aba^{-1}b^{-1} \in \Gamma(x_i)$. Similarly, $aba^{-1}b^{-1} \in \Gamma(x_i)$ for $0 \leq i \leq m$. Consequently $\Sigma \subseteq \Gamma(x_0,\ldots,x_{3m})$.

By Lemma 7.3, $\Gamma(x_{m-1},x_m)$ is a nontrivial 2-group and so $\Sigma_1 \neq 1$. Let $a \neq 1$ be an element in Σ_1 and let $u \in N(x_0) - \{x_1\}$. By Lemma 7.4, $a \in \Gamma(u,x_0,\ldots,x_{2m})$. By Lemma 7.2, $\Gamma(u,x_0,\ldots,x_{2m},x_{2m+1}) = 1$. Hence $a \notin \Gamma(x_{2m+1})$. Now suppose $\Sigma = 1$. Then a centralizes Σ_2. Let $v \in N(x_{3m}) - \{x_{3m-1}\}$. By Lemma 7.4 again, $\Sigma_2 \subseteq \Gamma(x_{2m},\ldots,x_{3m},v)$ and therefore $\Sigma_2 = a\Sigma_2a^{-1} \subseteq \Gamma(a(x_{2m}),\ldots,a(x_{3m}),a(v))$ and thus

$$\Sigma_2 \subseteq \Gamma(v,x_{3m},x_{3m-1},\ldots,x_{2m},a(x_{2m+1}),a(x_{2m+2}),\ldots,a(x_{3m}),a(v)).$$

Since $(v,x_{3m},x_{3m-1},\ldots,x_{2m},a(x_{2m+1}),a(x_{2m+2}),\ldots,a(x_{3m}),a(v))$ is an s-path, $\Sigma_2 = 1$, which is a contradiction. Hence $\Sigma \neq 1$. Finally, since $\Sigma \subseteq \Gamma(x_0,\ldots,x_{3m})$, it follows from Lemma 7.1 that $3m < s - 1$, i.e. $s < 4$. //

From now on, we assume that s is odd and we let $n = (s - 3)/2$.

Lemma 7.6 Let $xy \in E$. Then for each $(n+2)$-path $(x_0,x_1,\ldots,x_{n+2})$ in G, where $\{x_0,x_1\} = \{x,y\}$, $Z(\Gamma(x,y)) \not\subseteq \Gamma(x_0,\ldots,x_{n+2})$.

Proof. Let $\Gamma\{x,y\}$ be the stabilizer of $\{x,y\}$. Then $\Gamma\{x,y\}$ acts transitively on the set of all $(n+2)$-paths of the form $(x_0,x_1,\ldots,x_{n+2})$. Now since $\Gamma(x,y) \subseteq \Gamma\{x,y\}$, for every $a \in Z(\Gamma(x,y))$, $\Gamma(a(x),a(y)) = a\Gamma(x,y)a^{-1} = \Gamma(x,y) \subseteq \Gamma\{x,y\}$. Hence $\{a(x),a(y)\} = \{x,y\}$, from which it follows that $Z(\Gamma(x,y)) \lhd \Gamma\{x,y\}$.

Next, suppose $Z(\Gamma(x,y))$ fixes $[x] = (x_0,x_1,\ldots,x_{n+2})$. Then for any $[x'] = (x_0',x_1',\ldots,x_{n+2}')$ where $\{x_0',x_1'\} = \{x,y\}$, there exists $b \in \Gamma\{x,y\}$ such that $b[x'] = [x]$. Since $g[x] = [x]$ and $gb = bg$, for every $g \in Z(\Gamma(x,y))$, $g[x'] = gb^{-1}[x] = b^{-1}g[x] = b^{-1}[x] = [x']$. Hence if $Z(\Gamma(x,y))$ fixes an $(n+2)$-path of the form $(x_0,x_1,\ldots,x_{n+2})$, then it fixes every such $(n+2)$-path and so it fixes an s-path $(x_{-n-1},x_n,\ldots,x_{-1},x_0,x_1,x_2,\ldots,x_{n+2})$. However, since $Z(\Gamma(x,y)) \neq 1$, this contradicts Lemma 7.2. Consequently $Z(\Gamma(x,y)) \not\subseteq \Gamma(x_0,x_1,\ldots,x_{n+2})$. //

Lemma 7.7 Let $(x_0,\ldots,x_{3n})$ be a 3n-path and let $a \in Z(\Gamma(x_{n-1},x_n))$ with $a \notin \Gamma(x_{n-1},x_n,\ldots,x_{2n+1})$. Then $[a,b] \in \Gamma(x_0,\ldots,x_{3n})$ for each $b \in Z(\Gamma(x_{2n},x_{2n+1}))$.

Proof. By Lemma 7.4, if $n < i < 2n$, then a and b fix x_i. If $0 < i < n$, then $\partial(b(x_i),x_n) = \partial(b(x_i),b(x_n)) = \partial(x_i,x_n) < n$. Hence, by Lemma 7.4, a fixes $b(x_i)$ and thus $[a,b]$ fixes x_i. (Note that $[a,b] = a^{-1}b^{-1}ab$, $a^{-1}b^{-1}ab(x_i) = a^{-1}b^{-1}b(x_i) = a^{-1}(x_i) = x_i$.) If $2n < i < 3n$, the same argument applies. Hence $[a,b] \in \Gamma(x_0,\ldots,x_{3n})$. //

Lemma 7.8 $s < 7$.

Proof. Let a be given as in Lemma 7.7. If $[a,b] = 1$ for every

$b \in Z(\Gamma(x_{2n},x_{2n+1}))$, then $Z(\Gamma(x_{2n},x_{2n+1})) = aZ(\Gamma(x_{2n},x_{2n+1}))a^{-1} = Z(\Gamma(x_{2n},a(x_{2n+1})))$. Let $(v_0,\ldots,v_{n+2})$ be an $(n+2)$-path with $(v_0,v_1,v_2) = (x_{2n+1},x_{2n},a(x_{2n+1}))$. (Note that $a(x_{2n+1}) \neq x_{2n+1}$ by Lemma 7.6.) Then $Z(\Gamma(x_{2n},a(x_{2n+1}))) \subseteq \Gamma(v_0,\ldots,v_{n+2})$ by Lemma 7.4. But $Z(\Gamma(x_{2n},x_{2n+1})) \not\subseteq \Gamma(v_0,\ldots,v_{n+2})$ by Lemma 7.6. This contradiction shows that $[a,b] \neq 1$ for any $b \in Z(\Gamma(x_{2n},x_{2n+1}))$. By Lemma 7.7, it follows that $\Gamma(x_0,\ldots,x_{3n}) \neq 1$ and hence $3n \leq s - 1$, from which we obtain $s \leq 7$. //

Lemma 7.9 Suppose $s = 7$. If $xy \in E$ and $f \in Z(\Gamma(x,y))$, then $f^2 = 1$. Also if $(w_1,\ldots,w_5)$ is any 4-path and $d \in \Gamma(w_1,\ldots,w_5)$, then $d^2 = 1$.

Proof. Let $(w_0,w_1,w_2,\ldots,w_6,w_7)$ be a 7-path where $w_3 = x$, $w_4 = y$. By Lemma 7.4, $f \in \Gamma(w_1,w_2,\ldots,w_6)$. Since G is cubic, $f^2 \in \Gamma(w_0,w_1,\ldots,w_7) = 1$. Hence $f^2 = 1$.

If $(w_1,\ldots,w_5)$ is any 4-path, let $(w_0,\ldots,w_7)$ be a 7-path extending $(w_1,\ldots,w_5)$. By Lemma 7.6, there exists $g \in Z(\Gamma(w_2,w_3))$ such that $dg \in \Gamma(w_1,\ldots,w_6)$. Again, since G is cubic, $(dg)^2 \in \Gamma(w_0,\ldots,w_7) = 1$. Hence $(dg)^2 = 1$. Finally, from $g \in Z(\Gamma(w_2,w_3))$, $g^2 = 1$, $d \in \Gamma(w_1,\ldots,w_5)$, and $gd = dg$, it follows that $d^2 = 1$. //

To complete the proof of Theorem 7.1, now we need only to show that $s \neq 7$. Suppose $s = 7$. Let $(x_0,\ldots,x_7)$ be a 7-path. Let $b \in Z(\Gamma(x_4,x_5))$ with $b \notin \Gamma(x_1)$ and $c \in Z(\Gamma(x_5,x_6))$ with $c \notin \Gamma(x_2)$. (Here we apply Lemma 7.6.) Choose $a \in Z(\Gamma(x_1,x_2))$ such that $a \notin \Gamma(z)$ where z is given as in Fig.3.3. By Lemma 7.4, we have $a(b(x_1)) = b(x_1)$. Now since

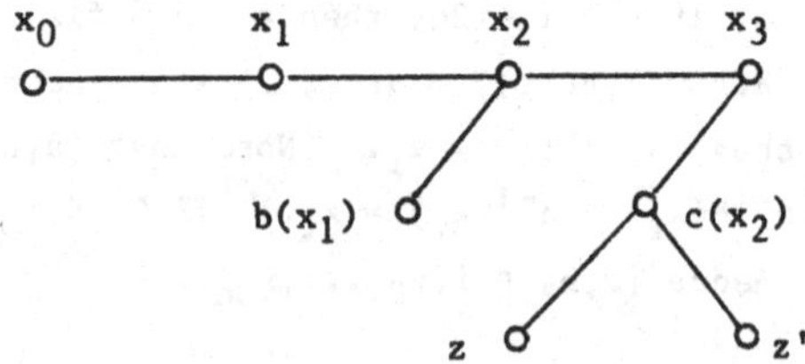

Figure 3.3

$N(c(x_2)) = \{x_3,z,z'\}$ and $c(x_3) = x_3$ (by Lemma 7.4), we also have

$c(b(x_1)) = z$ or z'. Next, by Lemma 7.4 again, a fixes x_3 and $c(x_2)$, and thus a switches z and z' (because $a \notin \Gamma(z)$). Hence d = cacab fixes x_1. ($d(x_1) = caca(b(x_1)) = cac(b(x_1)) = ca(z)$ or $ca(z')$. Suppose $c(b(x_1)) = z$. Then $d(x_1) = ca(z) = c(z') = x_1$ because $c(b(x_1)) = z$ and $c^2 = 1$. The case for $c(b(x_1)) = z'$ can be proved similarly.)

Next, by Lemma 7.4, $b \in \Gamma(x_2,x_3,x_4,x_5)$ and thus $d(x_2) = caca(b(x_2)) = caca(x_2) = ca(c(x_2)) = c(c(x_2)) = x_2$ because $c^2 = 1$ (by Lemma 7.9). Similarly, $d(x_3) = x_3$. Hence $d \in \Gamma(x_1,x_2,x_3,x_4,x_5)$ and so by Lemma 7.9, $d^2 = 1$. From d = cacab and the fact that $b^2 = 1$ and $a^2 = 1$, we obtain dba = cac. Hence $(dba)^2 = (cac)^2 = ca^2c = c^2 = 1$.

Finally, since $d \in \Gamma(x_1,x_2) \cap \Gamma(x_4,x_5)$, $b \in Z(\Gamma(x_4,x_5))$ and $a \in Z(\Gamma(x_1,x_2))$, we have $[d,b] = 1 = [d,a]$. Thus $d^2(ba)^2 = (dba)^2 = 1$ and so $(ba)^2 = 1 = [b,a]$. Hence $a \in Z(\Gamma(x_2))$, because $b \in \Gamma(x_2)$. Choose $g \in \Gamma(x_2)$ with $g(x_1) = x_3$. Since $a \in Z(\Gamma(x_1,x_2))$, by Lemma 7.4, we have $a \in \Gamma(y)$ for any vertex y such that $\partial(x_i,y) < 2$, $i = 1,2$. Hence $a = gag^{-1} \in gZ(\Gamma(x_1,x_2))g^{-1} = Z(\Gamma(g(x_1),g(x_2))) = Z(\Gamma(x_3,x_2)) = Z(\Gamma(x_2,x_3))$. However, by Lemma 7.4 again, $Z(\Gamma(x_2,x_3)) \subseteq \Gamma(z)$. Consequently $a \in \Gamma(z)$, contradicting the choice of a. //

Remarks.

1. Examples of 5-transitive and primitive 5-transitive cubic graphs can be found in Biggs [74;p.125]. An infinite family of strictly (primitive) 4-transitive cubic graphs has been constructed by Wong [67].

2. J. H. Conway has constructed an infinite family of connected, strictly 5-transitive cubic graphs (for details of construction, see Biggs [74;§19]).

3. The small 5-transitive cubic graphs known so far, in order of magnitude have 30, 90, 234, 468 and 650 vertices. Recently Biggs [82a] has constructed a 5-transitive cubic graph of order 2352. Biggs [82b] gives a construction of 5-transitive cubic graphs from 4-transitive ones which recovers Conway's result but the order of the graphs are smaller.

4. Gardiner [73] proved the following result which is similar to Tutte's theorem : Let G be an s-transitive graph with $s \geq 4$. Suppose deg G = p + 1, p an odd prime. Then s = 4, 5 or 7.

5. Weiss [73] also proved a similar result which is stated as follows : Let G be an s-transitive graph. If deg G = $1 + p^r n$ with p a prime and $r < p$, then $s \leq 7$ and $s \neq 6$; if deg G = $1 + p^r$, $r = 2^m$, $m \geq 1$, $p \neq 2$, then $s = 1$.

6. Further results on s-transitive cubic graphs can be found from Biggs [84], Biggs and Hoare [82], Biggs and Smith [71], Bouwer and Djoković [73], Djoković [71,73,74], Djoković and Miller [80], Gardiner [76b], Miller [71] and Weiss [77,78].

7. Using the classification of the finite simple groups, Weiss [81] proved the following : If G is a strictly s-transitive graph of degree $k > 2$, then $s \leq 5$ or $s = 7$.

8. A graph G is said to be <u>locally s-transitive</u> if the stabilizer of any vertex v acts regularly on the set of s-paths starting at v. Certain results on locally s-transitive graphs have been obtained by Bouwer [71] and Weiss [76b,79].

Exercise 3.7

1. Suppose G is a regular graph of degree at least three and the girth of G is γ. Prove that if G is strictly s-transitive, then $\gamma \geq 2s - 2$.

2. Let G = (V,E) be the incidence graph of the Fano plane which is depicted in Fig.3.4. (Here V consists of the seven points v_1, v_2, v_3, u_1, u_2, u_3, w and the seven lines $\ell_1 = \{v_1,u_3,v_2\}$, $\ell_2 = \{v_1,w,u_1\}$, $\ell_3 = \{v_1,u_2,v_3\}$, $\ell_4 = \{v_2,w,u_2\}$, $\ell_5 = \{v_2,u_1,v_3\}$, $\ell_6 = \{v_3,w,u_3\}$, $\ell_7 = \{u_1,u_2,u_3\}$; E consists of all the point-line pairs $\{P,\ell\}$ such that P lies on ℓ.)

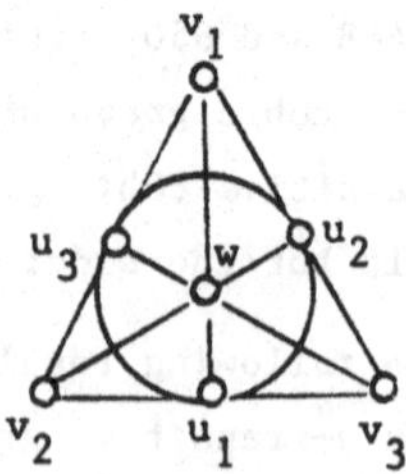

Figure 3.4.

Show that G is 4-transitive but not 5-transitive.

3. Prove that if G is a finite, connected, cubic graph which is strictly s-transitive, then for any $x \in V(G)$, $|\Gamma(x)| = 3 \times 2^{s-1}$ and $|\Gamma| = n \times 3 \times 2^{s-1}$ where $\Gamma = A(G)$.

4. Let $N = \{1,2,3,4,5,6\}$. Let S be the set of all 2-element subsets of N and T be the set of all partitions $ab|cd|ef$ of N into three 2-element sets. Suppose G is the graph with vertex set $V = S \cup T$ and edge set $\{s,t\}$ with $s \in S$, $t \in T$ such that s is one of the 2-element subsets making up t. Prove that G is 5-transitive.

5. Prove that the odd graphs O_k^* are strictly 3-transitive for all k > 3 (Biggs [74 ; p.118]).

6. Let G be a strictly s-transitive graph and let u, v be two adjacent vertices in G. Let Σ be the subgroup of A(G) which fixes u and v and all the vertices adjacent to either of them. Prove that Σ is a p-group for some prime p (Gardiner [73]).

8. 4-ultratransitive graphs

In the previous section we studied a class of highly symmetrical graphs. In this section we shall study another such class of graphs.

The girth, the diameter and the automorphism group of a graph G are denoted by $\gamma(G)$, $d(G)$ and Γ respectively. For a vertex x in G, we denote

$$N_i(x) = \{z \in G \mid \partial(x,z) = i\}.$$

Thus $N(x) = N_1(x)$ is the neighbourhood of x. If G is regular of degree (valency) k, we write deg G = k. The union of t disjoint copies of a graph H is denoted by tH and the complete t-paritite graph having r vertices in each partition is denoted by $O_r^t = O_r + O_r + \dots + O_r$ and thus O_r^2 is the complete bipartite graph $K_{r,r}$.

Suppose $U \subseteq V(G)$. Let $\langle U\rangle$ be the subgraph of G induced by U. A graph G is said to be ultrahomogeneous if every isomorphism from any induced subgraph $\langle U_1\rangle$ of G onto any induced subgraph $\langle U_2\rangle$ of G can be extended to an automorphism of G. If we restrict the above induced subgraphs $\langle U_1\rangle$ and $\langle U_2\rangle$ to the case $|U_1| = |U_2| \leq k$ for some fixed positive integer k, then G is said to be <u>k-ultratransitive</u>. Hence if G is ultrahomogeneous, then it is k-ultratransitive for every k = 1, 2, ..., $|G|$. If G is k-ultratransitive for some k > 1, then it is (k-1)-ultratransitive. A vertex-transitive graph is 1-ultratransitive. A graph G is <u>distance-transitive</u> if for any x, y, u, v ε V(G) such that $\partial(x,y) = \partial(u,v)$, there exists $\psi \;\varepsilon\; \Gamma$ such that $u = \psi(x)$ and $v = \psi(y)$.

Sheehan [74] and Gardiner [76a] characterized the class of finite ultrahomogeneous graphs. Their theorem is as follows.

Theorem 8.1 <u>If</u> G <u>is a finite ultrahomogeneous graph</u>, <u>then</u> G <u>is one of the following</u> :

(i) tK_r <u>for some</u> $t \geq 2$, $r \geq 1$;

(ii) O_r^t <u>for some</u> $t \geq 2$, $r \geq 1$;

(iii) C_5;

(iv) $L(K_{3,3})$, <u>the line graph of</u> $K_{3,3}$.

Cameron [80] and Y. Gol'fand (see Cameron [83]) characterized the class of finite 5-ultratransitive graphs. This class of graphs coincides with the class of graphs given in Theorem 8.1 and thus the result of Cameron and Gol'fand generalizes the result of Sheehan and Gardiner.

We know that nontrivial 3-ultratransitive graphs are <u>rank-3 graphs</u> (i.e. vertex-transitive graphs G which are not complete or null, such that for each vertex x of G, the orbits of $\Gamma(x)$ are $\{x\}$, $N(x)$ and $V(G) - (N(x) \cup \{x\})$). The class of rank-3 graphs have been extensively studied by many people, for instance, Hestenes [73], Hestenes and Higman [71], Hubaut [75], and Smith [75a,75b] etc. From their results, we can see that it is hard to characterize the class of finite 3-ultratransitive graphs. The main purpose of this section is to give a partial characterization of the class of finite 4-ultratransitive graphs. These results are due to Cameron [80] and Yap [83].

The following theorem is not difficult to prove and we shall leave it as an exercise.

Theorem 8.2 _Suppose_ G _is a_ 2-_ultratransitive graph_. _Then_

(i) G _is distance-transitive_;

(ii) _if_ G _is connected, then_ $d(G) \leq 2$;

(iii) _if_ G _is not connected, then_ $G = tK_r$, $t \geq 2$, $r \geq 1$;

(iv) _if_ G _is connected and bipartite, then_ $G = K_{r,r}$, $r \geq 1$.

The following theorem completely characterizes a special class of finite 4-ultratransitive graphs. This theorem generalizes the corresponding result of Theorem 8.1.

Theorem 8.3 _Suppose_ G _is a finite_ 4-_ultratransitive graph_. _If_ G _is connected, not bipartite and_ $\gamma(G) > 3$, _then_ $G = C_5$.

Proof. Let $k = \deg G$, $x \in V(G)$ and $N(x) = \{x_1,x_2,\ldots,x_k\}$. Let $y \in V(G)$ be such that $\partial(x,y) = 2$ and let $W = N(x) \cap N(y) = \{x_1,x_2,\ldots,x_c\}$. If $k = c$, it is easy to show that $G = K_{c,c}$ which is excluded by the hypothesis. Hence $k > c$. Since $\gamma(G) > 3$, for every $y_1 \in N(y) - W$, W and $N(y_1) \cap N(x)$ are disjoint subsets of $N(x)$, each of cardinality c. Hence

$$k \geq 2c \tag{1}$$

Let $m = |N_2(x)|$. Since $N_2(x)$ and $N(x)$ form two orbits of G under the action of $\Gamma(x)$, by counting the number of edges joining the vertices in $N_2(x)$ with the vertices in $N(x)$ in two different ways, we have

$$mc = k(k-1) \tag{2}$$

We shall first prove that $c = 1$.

Suppose $c \geq 2$. Let $B = N(x_1) \cap N(x_2) \cap N(x_3)$ and let $b = |B|$. Since $x \in B$, $b \geq 1$.

Let $W_1 = N(x_2) \cap N(x_3) - B$, $W_2 = N(x_3) \cap N(x_1) - B$ and $W_3 = N(x_1) \cap N(x_2) - B$. Then $|W_1| = |W_2| = |W_3| = c - b$. Let $X_1 = N(x_1) - (W_2 \cup W_3 \cup B)$, $X_2 = N(x_2) - (W_3 \cup W_1 \cup B)$, $X_3 = N(x_3) - (W_1 \cup W_2 \cup B)$, $\Delta = \{x_1,x_2,x_3\}$ and $U = V(G) - (N(x_1) \cup N(x_2) \cup N(x_3) \cup \Delta)$ (see Fig.3.5).

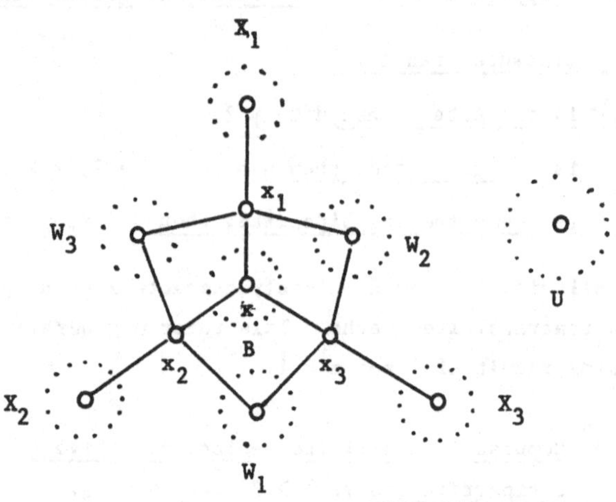

Figure 3.5

Then $|X_1| = |X_2| = |X_3| = k - \{2(c - b) + b\} = k + b - 2c \neq 0$. Let $z \in X_3$. Then B and $N(z) \cap N(x_1) \cap N(x_2)$ are disjoint subsets of $N(x_1) \cap N(x_2)$, each of cardinality b. Hence

$$c \geq 2b \tag{3}$$

It is clear that Δ, B, W_1, W_2, W_3, X_1, X_2, X_3 and U are disjoint subsets of V(G). Hence

$$\begin{aligned} |U| &= (m + k + 1) - \{3 + b + 3(c - b) + 3(k + b - 2c)\} \\ &= m - 2k + 3c - b - 2 \end{aligned} \tag{4}$$

Next since $x_4 \in U$, $U \neq \phi$. Now B and U form two orbits of G under the action of $\Gamma(\Delta)$. Hence, by counting the number of edges joining the vertices in U with the vertices in B in two different ways, we have

$$b(k - 3) = |U|d \tag{5}$$

where $d \geq 1$ is the number of vertices in B adjacent to a vertex $u \in U$. Hence

$$b(k - 3) \geq m - 2k + 3c - b - 2 \tag{6}$$

We note that W_1 and X_1 form two orbits of G under the action of $\Gamma(x_1,x_2,x_3)$. Let $w_1 \in W_1$. Then since $\gamma(G) > 3$, $N(x_1) \cap N(w_1) \subseteq X_1$ and since $|N(x_1) \cap N(w_1)| = c$, each vertex in W_1 is adjacent to c vertices in X_1. Let $x_1' \in X_1$. Then $N(x_1') \cap N(x_2) \cap N(x_3) \subseteq W_1$ and since $|N(x_1') \cap N(x_2) \cap N(x_3)| = b$, each vertex in X_1 is adjacent to b vertices in W_1. Hence

$$c(c - b) = b(k + b - 2c) \tag{7}$$

from which it follows that

$$c^2 = b(k + b - c) \tag{8}$$

Again, we note that W_1 and U form two orbits of G udner the action of $\Gamma(\Delta)$. Now since each vertex in W_1 is adjacent to $k - c - 2$ vertices in U and each vertex in U is adjacent to $b - d$ vertices in W_1, we have

$$(c - b)(k - c - 2) = |U|(b - d) \tag{9}$$

From (5) and (9), we obtain

$$c(k - c - 2) = b(|U| - c + 1) \tag{10}$$

From (8) and (10), we obtain

$$(k - c + b)(k - c - 2) = c(|U| - c + 1) \tag{11}$$

Substituting (4) into (11) and using (2), we obtain

$$(b - 1)k = c^2 - 3c + 2b \tag{12}$$

Substituting (8) into (12), we get

$$k = bc + 3c - b^2 - 2b \tag{13}$$

Substituting (13) back into (8), we get

$$c^2 - (b^2 + 2b)c + b^2 + b^3 = 0 \tag{14}$$

from which we obtain (since $c > 2b$)

$$c = b^2 + b \tag{15}$$

Hence, from (13), (15) and (2), we have

$$k = b^3 + 3b^2 + b \tag{16}$$

and

$$m = b^4 + 5b^3 + 6b^2 - b - 1 \tag{17}$$

Thus

$$b(k - 3) = b^4 + 3b^3 + b^2 - 3b$$

and

$$m - 2k + 3c - b - 2 = b^4 + 3b^3 + 3b^2 - b - 3$$

which contradicts (6). Consequently, $c = 1$.

Finally, suppose $k > 3$. Let $z_1 \in V(G)$ be such that $z_1 (\neq x) \in N(x_2)$ and $z_1 \notin N(y)$ (The existence of z_1 is guaranteed by the fact that $|N(y) \cap N(x_2)| = c = 1$ and $k = \deg G > 3$.) and let $z (\neq x,y) \in N(x_1)$. Now $\langle y,z,x\rangle \cong O_3 \cong \langle y,z_1,x\rangle$ and thus there exists $\psi \in \Gamma(y,x)$ such that $\psi(z) = z_1$. But $\{x_1\} = N(x) \cap N(y)$, therefore $\psi(x_1) = x_1$. However, $z_1 = \psi(z) \in N(x_1)$ and so $|N(z_1) \cap N(x)| > 2$, contradicting the fact that $|N(z_1) \cap N(x)| = c = 1$. Hence $k = 2$ and $G = C_5$. //

We shall next prove the following theorems which characterize partially the class of finite, 4-ultratransitive, connected graphs G having $\gamma(G) = 3$.

Lemma 8.4 Suppose $V(G) = N(x) \cup N(y)$ for any edge xy in G. Then each component of the complementary graph $\bar{G}$ of G is complete. In addition, if G is regular, then $G = O_r^t$ for some positive integers r and t.

Proof. Suppose yz, $zx \in E(\bar{G})$ and $xy \notin E(\bar{G})$. Then $xy \in E(G)$ and $z \notin N(x) \cup N(y)$, which is a contradiction. Hence each component of $\bar{G}$ is complete and $G = O_r^t$ if G is regular. //

Theorem 8.5 If G is vertex-transitive and $V(G) = N(x) \cup N(y)$ for any edge xy in G, then $G = O_r^t$, which is k-ultratransitive for all $k = 1, 2, \ldots$.

Proof. The fact that $G = O_r^t$ follows from Lemma 8.4. The last assertion can be proved by direct verification. //

Theorem 8.6 Suppose G is connected, 4-ultratransitive and $\gamma(G) = 3$. If $H = \langle N(x)\rangle$, $x \in V(G)$, is not connected, then G is the Schläfli graph or the line-graph $L(K_{3,3})$ of $K_{3,3}$.

Proof. (The Schläfli graph will be defined after the proof of this theorem.) Since H is 3-ultratransitive under the action of $\Gamma(x)$. By Theorem 8.2, $H = tK_r$ for some integers $t \geq 2$, $r \geq 2$.

We first distinguish two cases.

Case 1. $r \geq 3$.

Let $z, w_1, w_2, \ldots \in N(x)$ together with x form a maximum clique $K \cong K_{r+1}$ in $N[x]$. Let y_1, y_2 belong to a maximum clique $K' \neq K$ in $N[w_2]$. Since $d(G) = 2$ and G is 4-ultratransitive, y_1 is adjacent to a vertex $y_3 \in N(w_1)$. Now $\langle x,z,y_1,y_2 \rangle \cong \langle x,z,y_1,y_2 \rangle$ and thus there exists $\psi \in \Gamma$ such that ψ fixes x, z and y_1 separately and $\psi(y_2) = y_3$. However, since ψ fixes x and z, ψ permutes the vertices in $K - \{x,z\}$. But it is clear that $\psi(y_2) \neq y_3$ because y_1 and y_2 belong to a maximum clique in $N[w_2]$ whereas y_1 and y_2 do not belong to any maximum clique in $N[w_i]$ for some $w_i \in K - \{x,z\}$ (see Fig.3.6(a)).

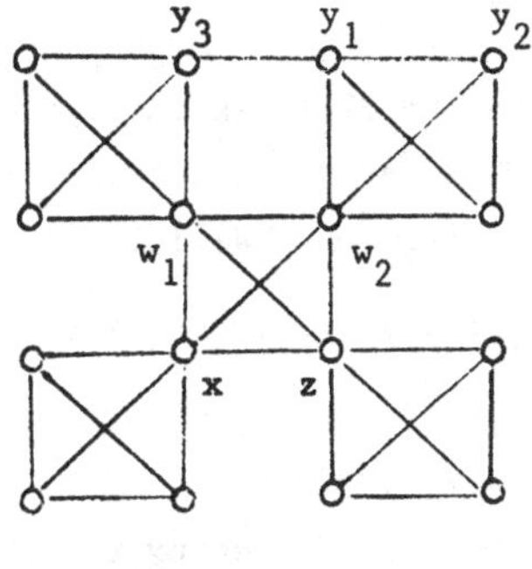

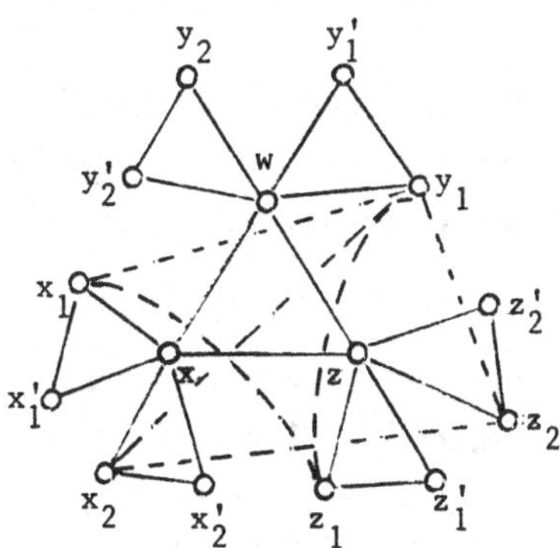

Figure 3.6

Case 2. $t \geq 3$, $r = 2$.

Let w, x and z form a triangle in G and let

$N(w) = \{x,z,y_1,y_1',\ldots\}$, $N(x) = \{z,w,x_1,x_1',\ldots\}$, $N(z) = \{w,x,z_1,z_1',\ldots\}$

where y_1y_1', x_1x_1', z_1z_1' $\ldots \in G$ (see Fig.3.6(b)).

It is clear that all the vertices in $Y = V(G) - (N(x) \cup N(z))$ are adjacent to w because they belong to the same orbit under the action of

$\Gamma(x,z)$. Now, since $d(G) = 2$, $\deg G = 2t$ and $H = tK_2$, and each edge of G is incident with exactly one triangle, without loss of generality, we can assume that

$$N(y_1) = \{w, y_1', x_1, z_1, x_2, z_2, \ldots\}$$

where x_1z_1, x_2z_2, ... ε G.

Next, it is clear that $A = \{y_2, y_2', \ldots, z_1', z_2', \ldots\}$ is such that each $a \in A$ is neither adjacent to x nor adjacent to y_1. Hence A forms an orbit under the action of $\Gamma(x,y_1)$. However, each vertex in $\{z_1', z_2', \ldots\}$ is adjacent to either y_2 or y_2' (but is not adjacent to both y_2 and y_2') and $\langle z_1', z_2', \ldots\rangle = O_{t-1}$. Hence

$$2 + (t - 1) = d_{\langle A\rangle}(y_2) + d_{\langle A\rangle}(y_2') = 2d_{\langle A\rangle}(z_1') = 2(t - 2)$$

which is true only if $t = 5$. However, when $t = 5$, G is the Schläfli graph which is 4-ultratransitive but not 5-ultratransitive.

Finally, for $t = 2$ and $r = 2$, it is not difficult to show that $G = L(K_{3,3})$ which is k-ultratransitive for any $k = 1,2,\ldots$. //

(The vertices of the Schläfli graph are the 27 lines in a general cubic surface, with two vertices adjacent whenever their corresponding lines meet. Suppose $G = (V,E)$ is the Schläfli graph. We let

$$V = \{x,y,z,x_1,\ldots,x_4,x_1',\ldots,x_4',y_1,\ldots,y_4,y_1',\ldots,y_4',z_1,\ldots,z_4,z_1',\ldots,z_4'\}$$

where x, y and z form a triangle and $x_1,\ldots,x_4,x_1',\ldots,x_4' \in N(x)$, $y_1,\ldots,y_4,y_1',\ldots,y_4' \in N(y)$, $z_1,\ldots,z_4,z_1',\ldots,z_4' \in N(z)$ $y_1y_1',\ldots,y_4y_4',x_1x_1',\ldots,x_4x_4',z_1z_1',\ldots,z_4z_4' \in G$. The adjaceny of the other vertices of G are as follows :

x_1: y_1, y_2, y_3, y_4, z_1, z_2', z_3', z_4'; x_2: y_1, y_2, y_3', y_4', z_1', z_2, z_3', z_4';

x_3: y_1, y_2', y_3, y_4', z_1', z_2', z_3, z_4'; x_4: y_1, y_2', y_3', y_4, z_1', z_2', z_3', z_4;

y_1: z_1, z_2, z_3, z_4; y_2: z_1', z_2', z_3, z_4; y_3: z_1', z_2, z_3', z_4;

y_4: z_1', z_2, z_3, z_4'.

Also x_1: y_1, y_2, y_3, y_4, z_1, z_2', z_3', $z_4' \Rightarrow x_1'$: y_1', y_2', y_3', y_4', z_1', z_2, z_3, z_4 and so on.)

Theorem 8.7 <u>Suppose</u> G <u>is connected</u>, 3-<u>ultratransitive and</u> $\gamma(G) = 3$. <u>If</u>

for each $x \in V(G)$, $H = \langle N(x) \rangle$ is connected and bipartite, then $G = O_r^3$ for some integer $r \geq 1$.

Proof. Since G is 3-ultratransitive, H is 2-ultratransitive under the action of $\Gamma(x)$. Thus, by Theorem 8.2, $H = K_{r,r}$ for some integer $r \geq 1$. It is clear that O_1^3 satisfies the hypothesis. Hence we may assume that $G \neq O_1^3$. Let $z \in N(x)$, $W = N(x) \cap N(z)$, $X = N(x) - (W \cup \{z\})$, and $Z = N(z) - (W \cup \{x\})$. Since $\gamma(G) = 3$, $W \neq \phi$. If $Z = \phi$, then G is complete and therefore $G = O_1^3$, which contradicts the assumption that $G \neq O_1^3$. Hence $Z \neq \phi$. Let $w \in W$, $z' \in Z$. Since $xw \in G$, $xz' \notin G$ and $\langle N(z) \rangle \cong K_{r,r}$, $wz' \in G$ and thus $Z = O_{r-1}$. By symmetry, $X = O_{r-1}$.

Suppose $N_2(x) \neq Z$. Then $Y = V(G) - (N(x) \cup N(z)) \neq \phi$. Since $d(G) = 2$, each $y \in Y$ is adjacent to some vertices in $W \cup Z$. If $yz' \in G$ for some $z' \in Z$, then $\langle N(z') \rangle \cong K_{r,r}$ together with $yz \notin G$ implies that $yw \in G$ for every $w \in W$. However, $\langle N(w) \rangle \cong K_{r,r}$ and $xz \in G$, imply that either $yx \in G$ or $yz \in G$ which is false. Hence $N_2(x) = Z$ and by Theorem 8.5, $G = O_r^t$ for some $t \geq 3$. But $t \nmid 4$. Hence $G = O_r^3$. //

Remarks.

1. According to Cameron [83], J. M. J. Buczak (in his D. Phil. thesis, Oxford University, 1980) has shown that, assuming the classification of the finite simple groups, the Schläfli graph and its complement are the only 4-ultratransitive graphs which are not 5-ultratransitive. Thus if we can prove, without using the classification of the finite simple groups, that if G is a finite, connected, 4-ultratransitive graph such that for any $x \in V(G)$, $H = \langle N(x) \rangle$ is connected and is not bipartite, then G is the complement of the Schläfli graph, we could have reconfirmed Buczak's result.

2. A graph G is homogeneous if whenever $U_1, U_2 \subseteq V(G)$ are such that $\langle U_1 \rangle \cong \langle U_2 \rangle$, then there exists an automorphism of G taking U_1 to U_2. A graph G is combinatorially homogeneous if whenever $U_1, U_2 \subseteq V(G)$ are such that $\langle U_1 \rangle \cong \langle U_2 \rangle$, then the number of vertices in G adjacent to all the vertices in U_1 and the number of vertices in G adjacent to all the vertices in U_2 are equal. Thus ultrahomogeneous graphs are homogeneous and homogeneous graphs are combinatorially homogeneous. Enomoto [81]

proved that every finite combinatorially homogeneous graph is ultrahomogeneous. From this result, it follows that homogeneous graphs are ultrahomogeneous, an earlier result proved by Ronse [78].

3. A graph G is said to be z-homogeneous if every isomorphism from any connected induced subgraph $\langle U_1 \rangle$ of G onto any induced subgraph $\langle U_2 \rangle$ of G can be extended to an automorphism of G. Weiss [76] determined all finite z-homogeneous graphs.

4. A connected graph G is said to be metrically k-transitive if, whenever two k-tuples $(x_1,\ldots,x_k)$ and $(y_1,\ldots,y_k)$ of vertices of G satisfy $\partial(x_i,x_j) = \partial(y_i,y_j)$ for all $i,j = 1,2,\ldots,k$, there is an automorphism ψ of G such that $\psi(x_i) = y_i$ for all $i = 1,2,\ldots,k$. Meredith [76] proved that a metrically 3-transitive graph of girth greater than 4 is a cycle and that a metrically 4-transitive graph of girth 4 is a complete bipartite graph or a complete bipartite graph minus a matching. Cameron [80] proved that the class of metrically 5-transitive graphs coincides with the class of ultrahomogeneous graphs given in Theorem 8.1 (except tK_r, $t \geq 2$, which is not a connected graph) from which he deduced that a metrically 6-transitive graph is one of the following : a complete multipartite graph, a complete bipartite graph minus a matching, a cycle, $L(K_{3,3})$, or the graph whose vertices are the 3-element subsets of a set consisting 6 elements, where two vertices are adjacent whenever the intersection of their corresponding subsets consists of 2 elements.

5. The definition of k-ultratransitivity suggests a number of variations which have been pointed out by Cameron [83]. For a detail account of the variations and for many interesting problems posed by Cameron, see Cameron [83].

6. The following are some other papers written on this subject : Cameron [77,79], Gardiner [78] and Woodrow [79].

Exercise 3.8

1. Prove that there does not exist any finite 3-ultratransitive tournament of order at least 4.
(This was first proved, using group theoretic methods, by W. M. Kantor in 1969. An elementary proof of this result was

subsequently given by Yap [83]. It has been proved that a finite tournament is 2-ultratransitive if and only if it is a Paley tournament, see Berggren [72a] and Kantor [69].)

2. Prove Theorem 8.2.

3. Prove that the graphs O_r^t and $L(K_{3,3})$ are ultrahomogeneous.

4. Prove that the following graph is 3-ultratransitive but not 4-ultratransitive.

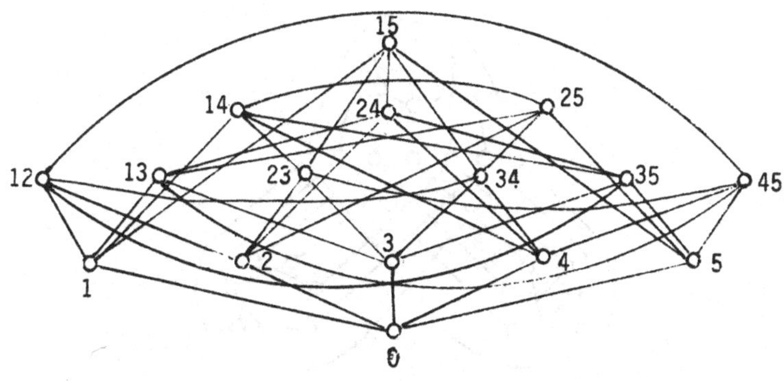

Figure 3.7

(This graph is connected, not bipartite and having girth greater than 3. This example indicates that probably there are quite a few 3-ultratransitive graphs which are connected, not bipartite, and having girth greater than 3.)

5. Prove that the Schläfli graph is 4-ultratransitive (Smith [75]).

6. Prove that there does not exist a 2-ultratransitive graph G such that for each $x \in V(G)$, $\langle N(x) \rangle \cong C_n$ where $n > 5$.

7. A graph G is <u>locally ultrahomogeneous</u> if whenever $U \subseteq V(G)$, then every automorphism of $\langle U \rangle$ extends to an automorphism of G. Prove that every locally ultrahomogeneous graph of finite order is ultrahomogeneous (Gardiner [76c]).

9. Hamilton cycles in Cayley graphs

In 1970 Lovász posed the question of whether or not every connected vertex-transitive graph has a Hamilton path (a Hamilton path will be abbreviated as an H-path). Up to now only four connected vertex-transitive graphs are known to have H-paths but do not have Hamilton cycles. These four graphs are the Petersen graph, the Coxeter graph (see Fig.3.8) and the graphs obtained from each of these two graphs by replacing each vertex with a triangle and joining the vertices as indicated in Fig.3.9. For this construction, we shall simply say that "each vertex is replaced by a K_3".

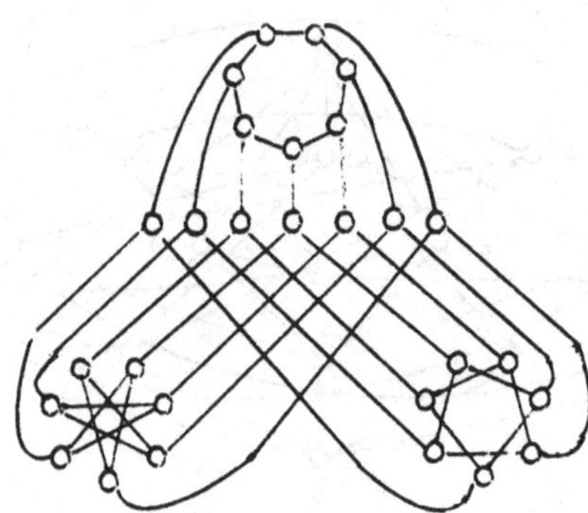

Figure 3.8 : The Coxeter graph

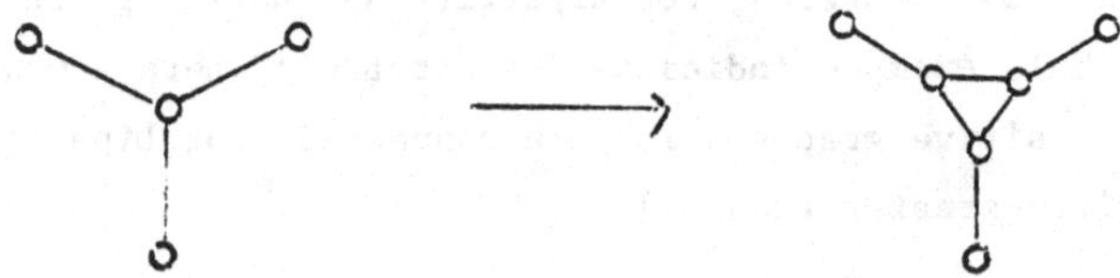

Figure 3.9 : "Replacing a vertex of valency 3 by a K_3"

These four graphs are not Cayley graphs. Thus it is interesting to know whether every connected Cayley graph is Hamiltonian. In fact, many people believe that the answer to this question is affirmative.

In this section we shall discuss some progress made towards the resolution of these two problems.

We recall that the β-product $B = P_m \times P_n$ of two paths P_m and P_n is the graph having $V(B) = \{(i,j) \mid 0 \leq i \leq m-1, 0 \leq j \leq n-1\}$ such that two vertices (i,j) and (h,k) of B are adjacent in B if and only if either $i = h$ and $|j - k| = 1$ or $j = k$ and $|i - h| = 1$. For each vertex $x = (i,j)$ in B we shall colour it with a blue colour if $i + j$ is even and we shall colour it with a red colour if $i + j$ is odd. A vertex of B coloured with a blue (resp. red) colour is called a _blue_ (resp. _red_) _vertex_. An edge joining two red (resp. blue) vertices of B is a _red edge_ (resp. _blue edge_) and an edge joining a red vertex and a blue vertex is a _purple edge_. We call a vertex x a _corner-vertex_ of B if its valency is 2; a _side-vertex_ if its valency is 3; and an _interior-vertex_ if its valency is 4.

The results of this section are due to Chen and Quimpo [81,83]. We need the following lemmas.

Lemma 9.1 _Let_ $m, n \geq 3$ _be integers. If_ mn _is odd, then_ $B = P_m \times P_n$ _has an_ H-_path from any blue vertex_ x _to any blue vertex_ $y \neq x$.

Proof. It is easy to prove, by induction on n, that Lemma 9.1 is true for $B = P_3 \times P_n$ where $n \geq 3$ is any odd integer (Ex.3.9(3)).

If $m, n \geq 5$, let $B_0 = \{(s,t) \in B \mid s \leq m - 3, t \leq n - 3\}$ and let $B_1 = B - B_0$. By symmetry, we can always assume that $x \in B_0$.

We consider two cases.

Case 1. $y \in B_0$.

By induction, B_0 has an H-path P from x to y. It is clear that B contains a side-edge e as shown in Fig.3.10(a). We can now replace e by an H-path in B_1 to obtain an H-path of B as depicted in Fig.3.10(a).

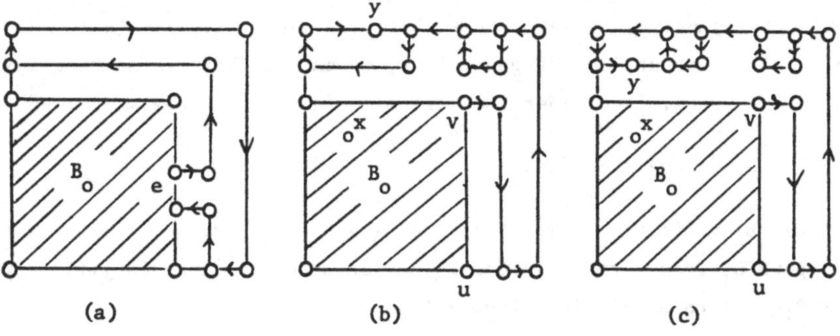

Figure 3.10

Case 2. $y \in B_1$.

By symmetry, we may assume that y lies on the upper part of B_1. Now B_0 has an H-path P from x to the upper-right corner vertex v. This path P can be extended to an H-path of B as depicted in Figs.3.10(b) and 3.10(c). (If $x = v$, then B_0 has an H-path P from x to the lower-right corner vertex u. This path can also be extended to an H-path of B.) //

Lemma 9.2 _If_ mn _is even, then_ $B = P_m \times P_n$ _has an_ H-_path from any corner-vertext_ x _to any vertex in_ B _whose colour is different from that of_ x.

Proof. By symmetry, we may take $x = (0,0)$ which is coloured blue. It is easy to prove by induction that Lemma 9.2 is true for $m = 2$ and any integer $n \geq 2$; and for $m = 3$ and any even integer $n \geq 2$ (Ex.3.9(4)). Using the method as described in the proof of Lemma 9.1, we can prove Lemma 9.2 for the general case $m \geq 4$, m even (without loss of generality) and any $n \geq 4$. //

Corollary 9.3 _If_ mn _is even_ ($m, n \geq 2$), _then_ $B = P_m \times P_n$ _has an_ H-_path connecting any two adjacent side-vertices._

We next prove

Lemma 9.4 _If_ mn _is even_ ($m, n \geq 4$), _then_ $B = P_m \times P_n$ _has an_ H-_path from any blue vertex_ x _to any red vertex_ y.

Proof. By symmetry, we may assume that m is even. Let $x = (i,j)$ and $y = (h,k)$. Without loss of generality, we may further assume that $i \leq h$ and $j \leq k$. We will consider the following four cases separately.

Case 1. $h = i$ and i is odd.

Let $B_2 = \{(s,t) \in B \mid t \leq j\}$ and $B_3 = B - B_2$. By Lemma 9.2, B_2 has an H-path Q_1 from x to $(0,j)$ and B_3 has an H-path Q_2 from $(0,j+1)$ to y. Thus Q_1Q_2 is an H-path in B from x to y.

Case 2. $h = i$ and i is even.

Suppose $0 < j$. We define B_2 and B_3 as in Case 1 above. Again, by Lemma 9.2, B_2 has an H-path Q_1 from x to $(m-1,j)$ and B_3 has an H-path Q_2 from $(m-1,j+1)$ to y. Then Q_1Q_2 is an H-path in B from x to y. On the other hand, if $j = 0$, we let $B_4 = \{(s,t) \in B \mid s \geq i, t \geq 1\}$ and $B_5 = \{(s,t) \in B \mid s < i\}$ (note that B_5 may be empty). By Lemma 9.2, B_4 has an H-path Q_1 from $(m-1,1)$ to y. The path Q_1 must contain an edge $(i,d)(i,d+1)$. By Corollary 9.3, B_5 has an H-path Q_2 from $(i-1,d)$ to $(i-1,d+1)$. Let Q_3 be the path obtained from Q_1 by replacing the edge $(i,d)(i,d+1)$ by Q_2. Then $(i,0)(i+1,0)\ldots(m-1,0)Q_3$ is an H-path in B from x to y.

Case 3. $h > i$ and i is odd.

Let $B_6 = \{(s,t) \in B \mid s \leq i\}$ and $B_7 = B - B_6$. By Lemma 9.2, B_6 has an H-path Q_1 from x to $(i,0)$ and B_7 has an H-path Q_2 from $(i+1,0)$ to y. Then Q_1Q_2 is an H-path in B from x to y.

Case 4. $h > i$ and i is even.

Let $B_8 = \{(s,t) \in B \mid s \geq i, t \geq j\}$ and $B_9 = \{(s,t) \in B \mid s \geq i, t < j\}$. By Lemma 9.2, B_8 has an H-path P from x to y. This path P must contain an edge $(c,j)(c+1,j)$ for some $c = i, i+1, \ldots, m-1$ or an edge $(i,d)(i,d+1)$ for some $d = 0, 1, \ldots, n-1$. Let this edge be $(c,j)(c+1,j)$, say. By Corollary 9.3, B_9 has an H-path Q from $(c,j-1)$ to $(c+1,j-1)$. Let R be the path obtained from P by replacing the edge $(c,j)(c+1,j)$ by Q. Then R is an H-path in $B_8 \cup B_9$ from x to y. It is easy to modify R to obtain an H-path in B from x to y. (Note that if $B_9 = \phi$, then P can be immediately modified to obtain an H-path in B from x to y.) This completes the proof of Lemma 9.4. //

A graph G is said to be <u>Hamilton-connected</u> if for any two vertices x, y of G, G has an H-path from x to y.

Lemma 9.5 <u>If</u> $m \geq 3$ <u>is odd and</u> $n \geq 2$, <u>then</u> $D = C_m \times P_n$ <u>is Hamilton-connected.</u>

Proof. We first prove that this lemma is true for $n = 2$. We note that $D = C_m \times P_2$ is obtained from two cycles C and C', each of length m, by joining corresponding vertices.

We need only to consider two cases.

Case 1. $x, y \in C$.

If x and y are adjacent in C, then this lemma is clearly true.

Suppose x and y are not adjacent in C. Since m is odd, one of the two semicycles of C joining x and y contains an odd number of vertices. We can now join x to y by an H-path as depicted in Fig.3.11(a).

Case 2. $x \in C$, $y \in C'$.

In this case, we can join x to y by an H-path as depicted in Fig.3.11(b).

The general case can be proved by induction on n (Ex.3.9(6)). //

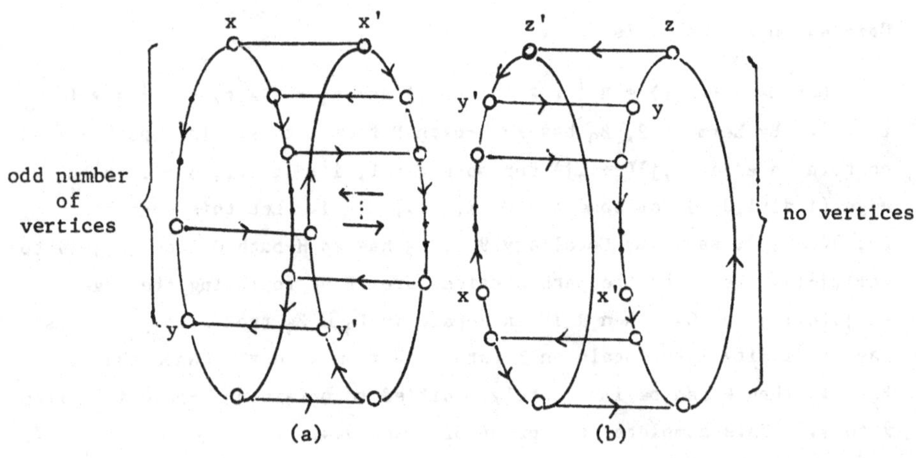

Figure 3.11

Corollary 9.6 <u>Suppose</u> $m > 3$ <u>and</u> $n > 1$ <u>are integers.</u> <u>Then each edge of</u> $D = C_m \times P_n$ <u>lies on a Hamilton cycle of</u> D.

Proof. This follows from Lemmas 9.4 and 9.5. //

Theorem 9.7 Suppose Γ is an abelian group of order at least 3 and the Cayley graph $G = G(\Gamma, S)$ is connected. Then for each $g \in S$, G contains a spanning subgraph isomorphic with $C_m \times P_n$ for some integers $m \geq 3$ and $n \geq 1$ such that the edge $(1,g)$ lies in $C_m \times P_n$.

Proof. Let $g = g_1$ and let $S' = \{g_1, \ldots, g_k\}$ be a minimal subset of S which generates Γ. Write $S_i = \{g_1, \ldots, g_i\}$ and let Γ_i be the subgroup of Γ generated by S_i. Then $(1,g_1)$ is an edge in $G_i = G(\Gamma_i, S_i \cup S_i^{-1})$ for all $i \geq 1$. If $\Gamma_1 = \Gamma$, then $G = G_1$ contains a spanning subgraph $C_m \times P_n$ where $m = |\Gamma| \geq 3$ and $n = 1$. On the other hand, if $\Gamma_1 \neq \Gamma$, then $g_2 \notin \Gamma_1$. Let n_2 be the smallest positive integer such that $g_2^{n_2} \in \Gamma_1$. It is clear that $\Gamma_2 = \Gamma_1 \cup g_2\Gamma_1 \cup \ldots \cup g_2^{n_2-1}\Gamma_1$. Moreover, the subgraph of G_2 induced by $g_2^j\Gamma_1$ is isomorphic with G_1 under the isomorphism $\psi : \Gamma_1 \to g_2^j \Gamma_1$ defined by $\psi(g) = g_2^j g$, $g \in \Gamma_1$. Also, since for each $h \in \Gamma_1$, $g_2^j h$ is adjacent to $g_2^{(j+1)}h$, G_2 contains a spanning subgraph isomorphic with $C_{n_1} \times P_{n_2}$ where $n_1 \geq 3$ is the order of g_1 such that $(1,g_1) \in E(C_{n_1} \times P_{n_2})$. (If $n_1 = 2$, then $G_2 \cong C_{n_2} \times P_2$ and $(1,g_1) \in E(C_{n_2} \times P_2)$ if $n_2 \geq 3$; or $G_2 = C_4 \times P_1$ if $n_2 = 2$.)

Now if $\Gamma_2 = \Gamma$, we have nothing more to prove. Otherwise by Corollary 9.6, G_2 contains a spanning cycle C_{m_2}, where $m_2 = n_1 n_2$, such that $(1,g_1) \in E(C_{m_2})$. By the same argument, G_3 contains a spanning subgraph $C_{m_2} \times P_{n_3}$ where n_3 is the smallest positive integer such that $g_3^{n_3} \in \Gamma_2$. The validity of Theorem 9.7 thus follows by induction. //

Theorem 9.8 For any edge (x,y) of a connected Cayley graph $G = G(\Gamma, S)$ of an abelian group Γ, G has a Hamilton cycles containing (x,y).

Proof. Since $(x,y) \in E(G)$, $x^{-1}y \in S$. By Theorem 9.7 and Corollary 9.6, G contains a Hamilton cycle containing the edge $(1,x^{-1}y)$. Now the map $\psi : \Gamma \to \Gamma$ defined by $\psi(h) = xh$ for every $h \in \Gamma$ is an automorphism of G such that $\psi : (1,x^{-1}y) \to (x,y)$. Thus (x,y) lies on a Hamilton cycle of G. //

Remarks.

1. Let p be a prime. Theorem 5.1 shows that every vertex-transitive graph of order $p \geq 3$ is Hamiltonian. Alspach [79] proved that if G

is a connected vertex-transitive graph of order 2p, then G is Hamiltonian. Marušič (according to Alspach [81]) has proved that if G is a connected vertex-transitive graph of order 3p, then G is Hamiltonian. Marušič and Parsons [83,81] have proved that if G is a connected vertex-transitive graph of order 4p or 5p, then G has an H-path.

2. Lipman [85] proved that if G is a connected vertex-transitive graph such that A(G) has a nilpotent subgroup acting transitively on V(G), or if $|V(G)|$ is a prime power, then G has an H-path.
3. Chen and Quimpo [83] proved that every connected Cayley graph of order pq, where p and q are distinct primes, is Hamiltonian.
4. Further results on Hamilton cycles in Cayley graphs can be found in Holsztyński and Strube [78], Housman [81] and Witte [82].

Exercise 3.9

1. Prove that the graph obtained from the Petersen graph by "replacing each vertex by a K_3" is vertex-transitive.
2. Prove that the Coxeter graph is vertex-transitive but not a Cayley graph and that the graph obtained from the Coxeter graph by "replacing each vertex by a K_3" is also vertex-transitive but not a Cayley graph.
3. Prove that for any odd integer $n > 3$, $B = P_3 \times P_n$ has an H-path from any blue vertex x to any blue vertex $y \neq x$.
4. Prove that $B = P_2 \times P_n$ for any integer $n > 2$ or $B = P_3 \times P_n$ for any even integer $n > 2$, has an H-path from the blue vertex $x = (0,0)$ to any red vertex y. Prove Lemma 9.2 for the general case.
5. Prove Corollary 9.3.
6. Prove the general case of Lemma 9.5 by induction on n.
7. Show that if $m > 4$ is even and $n = 2$ or 3, then any blue vertex x is connected to any red vertex y by a Hamilton path in $C_m \times P_n$.
8. Let Γ be a group and let N, H be subgroups of Γ such that $N \lhd \Gamma$, $N \cap H = \{1\}$ and Γ is generated by $N \cup H$. Then Γ is said to be a <u>semidirect product</u> of N by H. Suppose Γ is a semidirect product of

a cyclic group of prime order by a finite abelian group of odd order. Prove that if the Cayley graph $G = G(\Gamma,S)$ is connected, then it is Hamiltonian (Marušič [83]).

9. Let $p \geq 5$ be a prime. Prove that any two vertices x and y in the Cayley graph $G(Z_p,S)$ are connected by an H-path (Alspach [79]).

10. Let $G = G(\Gamma,S)$ be a connected Cayley graph such that for each $g \in S$, $\langle S - \{g,g^{-1}\}\rangle \neq \Gamma$. Prove that if G is Hamiltonian, then any edge of G is contained in a Hamilton cycle of G.

11. Prove that a connected Cayley graph $G = G(\Gamma,S)$ of an abelian group Γ of order at least 3 is Hamilton-connected (i.e. any two distinct vertices of G are connected by an H-path of G) if it is not bipartite and $\deg G \geq 3$ (Chen and Quimpo [81]).

10. Concluding remarks

To conclude this chapter, we shall now briefly mention some other interesting and important results about symmetries of graphs which we have not been able to discuss in detail due to lack of space. There are no less than three hundred papers published on this subject. We refer the readers to the two excellent survey articles by Babai [81] and Cameron [83] for further information.

1. Further properties of vertex-transitive graphs

Certain general properties of VT-graphs have been studied. Watkins [70] proved that if G is a connected VT-graph such that the connectivity k(G) of G is less than deg G, then k(G)/deg G has the least upper bound 3/2, which is never attained. He also proved that for any connected edge-transitive graph G, k(G) = deg G. Mader [70] proved that if G is a connected VT-graph such that the density of G is at least 4, then k(G) = deg G. Watkin's and Mader's results were further exploited and/or generalized by Green [75]. Babai [78] investigated the minimum chromatic number of Cayley graphs. Babai and Frankl [78,79] studied the isomorphism problems of Cayley graphs. Godsil [81a] studied the connectivity of minimal Cayley graphs. Goldschmidt [80] studied the automorphism groups of cubic graphs. The automorphism groups of Cayley graphs have also been studied by several people, including Babai and Godsil [82], Godsil [80,83] and Imrich and Watkins [76]. Lorimer [83] described certain methods for constructing VT-graphs of degree 3. Lorimer [84b] proved that there are exactly 6 cubic graphs G of order at most 120 which are not bipartite and on which no automorphism group of G acts regularly. Coxeter, Frucht and Powers [81] have constructed 350 cubic VT-graphs G of order at most 120 such that A(G) acts regularly on V(G). Yap [70] showed how to construct Cayley graphs G(Γ,S) of abelian groups Γ in which Γ has maximum number of _internal stable sets_ (subsets $S \subseteq \Gamma$ such that the subgraph induced by S is a null graph).

2. Distance-transitive graphs

The following are some important results on distance-transitive graphs.

Theorem 10.1 (D. H. Smith [71]) _An imprimitive distance transitive graph is either bipartite or antipodal. (Both possibilities can occur in the same graph.)_

(An _imprimitive distance-transitive graph_ is a distance-transitive graph G such that the natural action of A(G) on V(G) is imprimitive. A graph of diameter d is said to be _antipodal_ if, when three vertices u, v, w are such that $\partial(u,v) = \partial(u,w) = d$, then either $\partial(v,w) = d$ or $v = w$.)

Theorem 10.2 (Biggs and Smith [71]) _There are precisely twelve distance-transitive cubic graphs. The orders of these graphs are_ 4, 6, 8, 10, 14, 18, 20, 20, 28, 30, 90 _and_ 102.

D. H. Smith [74] determined all (a finite number) the distance-transitive graphs of degree 4. Recently, the following important theorem is proved, using deep results about finite simple groups.

Theorem 10.3 (Cameron [82]) _There are only finitely many finite distance-transitive graphs of given degree_ $k > 2$.

For a detailed study on distance-transitive graphs, see Biggs [74; Chapters 20, 21 and 22]. The study of distance-transitive graphs is still active. The more recent papers include Cameron, Saxl and Seitz [83], and Gardiner [82].

3. Long cycles in vertex-transitive graphs

In §9, we have shown that several classes of connected Cayley graphs possess Hamilton cycles. However, the problems that whether every connected Cayley graph is Hamiltonian and whether every connected vertex-transitive graph contains a Hamilton path remain open. We now mention a result of this kind from another approach.

Theorem 10.4 (Babai [79]) Every connected vertex-transitive graph of order $n > 4$ has a cycle of length greater than $\sqrt{3n}$.

4. The characteristic polynomial of a graph

Suppose G is a graph with vertex set $V = \{v_1,\ldots,v_n\}$. The adjacency matrix $A[G] = (a_{ij})$ of G is a matrix such that $a_{ij} = 1$ if $v_iv_j \in E(G)$ and $a_{ij} = 0$ if $v_iv_j \notin E(G)$. There are relations between certain properties of G and the eigenvalues of A[G]. The following are some typical results of this kind.

Theorem 10.5 (Mowshowitz [69]; Petersdorf and Sachs [70]) For a graph G, if all eigenvalues of A[G] are distinct, then the automorphism group of G is an elementary abelian 2-group.

Theorem 10.6 (Cameron [83]) For a graph G, if all eigenvalues of A[G] have multiplicity at most 3, then the automorphism group of G is soluble.

For short proofs of the above results, see Cameron [83].

Theorem 10.7 (Yap [75]) The characteristic polynomial of the adjacency matrix A[M] of a multi-digraph (or weighted-digraph) M whose automorphism group is nontrivial is the product of the symmetric part of

M *and the characteristic polynomial of the complementary part of* M.

(For definitions of weighted-digraph, the symmetric part of M and the characteristic polynomial of the complementary part of M, see Yap [75].) Theorem 10.7 generalizes some results of Collatz and Sinogowitz [57], and Mowshowitz [73]. From Theorem 10.7, we have

Corollary 10.8 *If the characteristic polynomial of* A[G] *of a graph* G *is irreducible over the field* Q *of rational numbers, then the automorphism group of* G *is trivial.*

For further results about the relations between the eigenvalues of A[G] and the properties of G, the readers may refer to the excellent book by Cvetković, Doobs and Sachs [80].

5. Graphs with given constant link

Suppose G is a vertex-transitive graph. Then for any two vertices x and y of G, the subgraphs induced by the neighbourhoods of x and y are isomorphic. It is interesting to know whether for a given graph L, there exist graphs G such that $\langle N(x)\rangle \cong L$ for any $x \in V(G)$. If such a graph G exists, then L is called a *link graph*. Several papers on this subject have been published. Some of these papers investigated which graphs are link graphs and some characterized all graphs with a given constant link. For instance, Hall [80] proved that the Petersen graph P is a link graph and that there are exactly three isomorphism classes of connected graphs with constant link P. Other papers on this subject can be found from the references of Vogler [84].

6. Vertex-transitive graphs of small order

Using the results of §5, we can construct all VT-graphs of prime order p. Yap [73] constructed all VT-graphs G of orders 4, 6, 8, 10 and 12 (except deg G = 5) by hand. Using computer, McKay [79] listed all the 1031 VT-graphs of order less than 20, together with many of their properties.

7. Symmetric graphs characterized by their local properties

In the exploratory paper by Praeger [82a], many results by Cameron [72,74], Praeger [82b] and others on symmetric graphs characterized by their local properties are surveyed and many interesting problems are posed.

REFERENCES

B. Alspach, Point-symmetric graphs and digraphs of prime order and transitive permutation groups of prime degree, J. Combin. Theory, Ser.B, 15 (1973), 12-17.

————, On constructing the graphs with a given permutation group as their group, Proc. of the Fifth Southeastern Conf. on Combinatorics, Graph Theory and Computing, Congr. Numer. X (1974), 187-207.

————, Hamilton cycles in vertex-transitive graphs of order 2p, Proc. of the Tenth Southeastern Conf. on Combinatorics, Graph Theory and Computing, Congr. Numer. XXIII (1979), 131-139.

————, The search for long paths and cycles in vertex-transitive graphs and digraphs, Combinatorial Math. VIII (Ed. K. L. McAvaney), Proc. of The Eight Australian Conf. on Combinatorial Mathematics 1980, Springer-Verlag Lecture Notes in Math. 884 (1981), 14-22.

B. Alspach and T. D. Parsons, A construction for vertex-transitive graphs, Canad. J. Math., vol.34, no.2 (1982), 307-318.

B. Alspach and R. J. Sutcliffe, Vertex-transitive graphs of order 2p, Second Internat. Conf. on Combinatorial Mathematics, Annals N. Y. Acad. Sci. 319 (1979), 19-27.

L. Babai, On the minimum order of graphs with given group, Canad. Math. Bull. 17 (1974), 467-470.

————, Chromatic number and subgraphs of Cayley graphs, Theory and Applications of Graphs (Eds. Y. Alavi and D. R. Lick; Proc. Internat. Conf., Western Mich. Univ., Kalamazoo, Mich., 1976), Springer-Verlag Lecture Notes in Math. 642 (1978), 10-22.

————, Long cycles in vertex-transitive graphs, J. Graph Theory 3, no.3 (1979), 301-305.

————, On the abstract group of automorphisms, Combinatorics (Ed. H. N. V. Temperley), London Math. Soc. Lecture Note Series 52 (1981), 1-40.

L. Babai and P. Frankl, Isomorphisms of Cayley graphs I, Combinatorics (Keszthely, 1976), Colloq. Math. Jànos Bolyai 18 (1978), 35-52.

———, Isomorphisms of Cayley graphs, Acta Mathematica, vol.34, nos. 1-2 (1979), 177-184.

L. Babai and C. D. Godsil, On the automorphism groups of almost all Cayley graphs, Europ. J. Combinatorics 3 (1982), 9-15.

G. Baron, Über asymmetrische Graphen, Math. Nachr. 46 (1970), 25-46.

G. Baron and W. Imrich, Asymmetrische reguläre Graphen, Acta Math. Acad. Sci. Hungar. 20 (1969), 135-142.

M. Behzad and G. Chartrand, Introduction to the Theory of Graphs, Allyn and Bacon Inc., Boston, 1971.

M. Behzad, G. Chartrand and L. Lesniak-Foster, Graphs and Digraphs, Wadsworth International Mathematical Series, 1979.

J. L. Berggren, An algebraic characterization of finite symmetric tournaments, Bull. Austral. Math. Soc. 6 (1972), 53-59.

———, An algebraic characterization of symmetric graphs with p points, p an odd prime, Bull. Austral. Math. Soc. 7 (1972), 131-134.

N. L. Biggs, Algebraic Graph Theory, Cambridge Tracts in Mathematics, vol.67, 1974.

———, Aspects of symmetry in graphs, Algebraic Methods in Graph Theory (Eds. L. Lovàsz and V. T. Sòs), Colloq. Math. Soc. Jànos Bolyai 25 (1981), 27-35.

———, A new 5-arc-transitive cubic graph, J. Graph Theory 6, no.4 (1982), 447-451.

———, Constructing 5-arc transitive cubic graphs, J. London Math. Soc., Ser.2, 26, no.2 (1982), 193-200.

N. L. Biggs and M. J. Hoare, The sextet construction for cubic graphs, Combinatorica, vol.3, no.2 (1983), 153-165.

N. L. Biggs and D. H. Smith, On trivalent graphs, Bull. London Math. Soc. 3 (1971), 155-158.

N. L. Biggs and A. T. White, Permutation Groups and Combinatorial Structures, London Math. Soc. Lecture Note Series 33, 1979.

Z. Bohdan, Involutory pairs of vertices in transitive graphs, Mat. Časopis Solven. Akad. Vied 25, no.3, (1975), 221-222.

B. Bollobàs, The asymptotic number of unlabelled regular graphs, J. London Math. Soc., Ser.2, 26 (1982), 201-206.

J. A. Bondy and U. S. R. Murty, Graph Theory with Applications, MacMillan Press, 1976.

I. Z. Bouwer, Vertex and edge transitive, but not 1-transitive graphs, Canad. Math. Bull. 13 (1970), 231-237.

———, On locally s-regular graphs, Proc. of the Twenty-Fifth Summer Meeting of the Canad. Math. Congress, Lakehead Univ., Thunder Bay, Ont., (1971), 293-302.

———, On edge but not vertex transitive regular graphs, J. Combin. Theory, Ser.B, 12 (1972), 32-40.

I. Z. Bouwer and D. Z. Djoković, On regular graphs III, J. Combin. Theory, Ser.B, 14 (1973), 268-277.

P. J. Cameron, Permutation groups with multiply transitive suborbits, Proc. London. Math. Soc. (3), 25 (1972), 427-440.

———, Suborbits in transitive permutation groups, Combinatorial Group Theory (Eds. M. Hall Jr. and J. H. van Lint), Math. Centre Tracts 57 (1974), 98-129.

———, A note on triple transitive graphs, J. London Math. Soc., Ser.2, 15 (1977), 197-198.

———, Multiple transitivity in graphs, Graph Theory and Combinatorics (Ed. R. J. Wilson), Research Notes in Math. 34, Pitman, London (1979), 38-48.

———, 6-transitive graphs, J. Combin. Theory, Ser.B, 28 (1980), 168-179.

———, There are only finitely many distance-transitive graphs of given valency greater than two, Combinatorica 2, no.1 (1982), 9-13.

———, Automorphism groups of graphs, Selected Topics in Graph Theory II (Eds. L. W. Beineke and R. J. Wilson), Academic Press (1983), 89-127.

P. J. Cameron, C. E. Praeger, J. Saxl and G. M. Seitz, On the Sims' conjecture and distance-transitive graphs, Bull. London Math. Soc. 15 (1983), 499-506.

A. Cayley, On the theory of groups, Proc. London Math. Soc. 9 (1878), 126-133.

C. Y. Chao, On a theorem of Sabidussi, Proc. Amer. Math. Soc. 15 (1964), 291-292.

———, On groups and graphs, Trans. Amer. Math. Soc. 111 (1965), 488-497.

———, On the classification of symmetric graphs with a prime number of vertices, Trans. Amer. Math. Soc. 158 (1971), 247-256.

C. Y. Chao and J. G. Wells, A class of vertex-transitive digraphs, J. Combin. Theory, Ser.B, 14 (1973), 246-255.

C. C. Chen and N. F. Quimpo, On some classes of Hamiltonian graphs, Southeast Asian Bull. Math. (special issue) (1979), 252-258.

———, On strongly Hamiltonian abelian group graphs, Combinatorial Math. VIII (Ed. K. L. McAvaney), Proc. of the Eight Australian Conf. on Combinatorial Mathematics 1980, Springer-Verlag Lecture Notes in Math. 884 (1981), 23-34.

———, Hamiltonian Cayley graphs of order pq, Combinatorial Math. X (Ed. L. R. A. Casse), Springer-Verlag Lecture Notes in Math. 1036 (1983), 1-5.

C. C. Chen and H. H. Teh, Constructions of point-colour-symmetric graphs, J. Combin. Theory, Ser.B, 27 (1979), 160-167.

L. Collatz and A. Sinogowitz, Spetkren endlicher Graphen, Abd. Math. Semi. Univ. Humburg 21 (1957), 63-77.

H. S. M. Coxeter, R. Frucht and D. L. Powers, Zero-Symmetric Graph : Trivalent Graphical Regular Representations of Groups, Academic Press, New York, London, 1981.

D. M. Cvetkovič, M. Doob and H. Sachs, Spretra of Graphs : Theory and Application, Academic Press, New York, London, 1980 (2nd Edition, VEB Deutscher Verlag der Wissenschaften, Berlin, 1982)

D. Z. Djokovič, On regular graphs I, J. Combin. Theory, Ser.B, 10 (1971), 253-263.

———, On regular graphs II, J. Combin. Theory, Ser.B, 12 (1972), 252-259.

———, On regular graphs IV, J. Combin. Theory, Ser.B, 15 (1973), 167-173.

———, Automorphisms of graphs and coverings, J. Combin. Theory, Ser.B, 16 (1974), 243-247.

D. Z. Djokovič and G. L. Miller, Regular groups of automorphisms of cubic graphs, J. Combin. Theory, Ser.B, 29 (1980), 195-230.

H. Enomoto, Combinatorially homogeneous graphs, J. Combin. Theory, Ser.B, 30 (1981), 215-223.

P. Erdös and A. Rényi, Asymmetric graphs, Acta Math. Sci. Hungar. 14 (1963), 295-315.

J. H. Folkman, Regular line-symmetric graphs, J. Combin. Theory 3 (1967), 215-232.

R. Frucht, Hertellung von Graphen mit vorgegebener abstrakten Gruppe, Composito Math. 6 (1938), 239-250.

———, Graphs of degree 3 with given abstract group, Canad. J. Math. 1 (1949), 365-378.

A. D. Gardiner, Arc-transitivity in graphs, Quart. J. Math. Oxford (2), 24 (1973), 399-407.

———, Arc-transitivity in graphs II, Quart. J. Math. 25 (1974), 163-167.

———, Homogeneous graphs, J. Combin. Theory, Ser.B, 20 (1976), 94-102.

———, Arc transitivity in graphs III, Quart. J. Math. Oxford (2) 27, (1976), 313-323.

———, Homogeneous graphs and stability, J. Austral. Math. Soc., Ser.A, 21 (1976), 371-375.

———, Homogeneity conditions in graphs, J. Combin. Theory, Ser.B, 24 (1978), 301-310.

———, Symmetry conditions in graphs, Surveys in Combinatorics, London Math. Soc. Lecture Note Series 38 (Ed. B. Bollobàs) (1979), 22-43.

———, Classifying distance-transitive graphs, Combinatorial Mathematics IX (Eds. E. J. Billington, S. Oates-Williams and A. P. Street), Springer-Verlag Lecture Notes in Math. 952 (1982), 67-88.

C. D. Godsil, GRR's for non-solvable groups, Colloq. Math. Soc. Jànos Bolyai 25 (1978), 221-239.

———, Neighbourhoods of transitive graphs and GRRs, J. Combin. Theory, Ser.B, 29, no.1, (1980), 116-140.

———, Connectivity of minimal Cayley graphs, Arch. Math. (Basel) 37 (1981), 473-476.

———, On the full automorphism group of a graph, Combinatorica 1 (1981), 243-256.

———, The automorphism groups of some cubic Cayley graphs, Europ. J. Combinatorics 4 (1983), 25-32.

D. M. Goldschmidt, Automorphisms of trivalent graphs, Annals Math. 111 (1980), 377-406.

A. C. Green, Structure of vertex-transitive graphs, J. Combin. Theory, Ser.B, 18 (1975), 1-11.

J. I. Hall, Locally Petersen graphs, J. Graph Theory 4 (1980), 173-187.

F. Harary, Graph Theory, Addison-Wesley, Reading, Mass., 1969.

F. Harary and E. M. Palmer, Graphical Enumeration, Academic Press, New York and London, 1973.

M. D. Hestenes, On the use of graphs in group theory, New Directions in the Theory of Graphs (Ed. F. Harary), (1973), 97-128.

M. D. Hestenes and D. G. Higman, Rank 3 groups and strongly regular graphs, Computers in Algebra and Number Theory, SIAM-AMS Proceedings 4 (1971), 141-159.

D. Hetzel, Über reguläre graphische Darstellungen von auflösbaren Gruppen, Diplomarbeity, Technische Universitat Berlin, 1976.

W. Holsztyński and R. F. E. Strube, Paths and circuits in finite groups, Discrete Math. 22 (1978), 263-272.

D. F. Holt, A graph which is edge transitive but not arc transitive, J. Graph Theory 5 (1981), 201-204.

D. A. Holton, The König question, Proceedings of the Fifth British Combinatorial Conference (Aberdeen, 1975), Congr. Numer. XV, (1976), 323-342.

D. Housman, Enumeration of Hamiltonian paths in Cayley diagrams, Aequationes Math. 23 (1981), 80-97.

X. L. Hubaut, Strongly regular graphs, Discrete Math. 13 (1975), 357-381.

W. Imrich, Graphen mit transitiver Automorphismengruppe, Monatsh. Math. 73 (1969), 341-347.

————, Graphs with transitive Abelian automorphism groups, Combinatorial Theory and its Applications II, Colloq. Soc. Jànos Bolyai 4, (1970), 651-656.

W. Imrich and M. E. Watkins, On automorphism groups of Cayley graphs, Period. Math. Hungar. 7 (1976), 243-258.

H. Izbicki, Graphen 3, 4 und 5 Grades mit vorgegebenen abstrakten Automorphismengruppen, Farbenzhlen und Zusamenhängen, Montsh. Math. 61 (1957), 42-50.

————, Unendliche Graphen endlichen Grades mit vorgegebenen Eigenschaften, Montash. Math. 63 (1959), 298-301.

————, Reguläre Graphen beliebigen Grades mit vorgegebenen Eigenschaften, Monatsh. Math. 64 (1960), 15-21.

————, Über asymmetrische reguläre Graphen, Monatsh. Math. 73 (1969), 112-125.

I. N. Kagno, Linear graphs of degree ≤ 6 and their groups, Amer. J. Math. 68 (1946), 505-520; 69 (1947), 872; 77 (1955), 392.

W. M. Kantor, Automorphism groups of designs, _Math. Z._ 109 (1960), 246-252.

M. H. Klin, On edge but not vertex transitive graphs, _Algebraic Methods in Graph Theory_ (Ed. L. Lovàsz and V. T. Sòs), _Colloq. Math. Soc. Janos Bolyai_ 25 (1981), 399-404.

D. König, _Theorie der endlichen und unendlichen Graphen_, Leipzig (1936); Reprinted by Chelsea, New York (1950).

C. K. Lim, On graphs with transitive Abelian automorphism groups, _Malaysian J. Science_ 4 (1976), 107-110.

M. J. Lipman, Hamiltonian cycles and paths in vertex-transitive graphs with abelian and nilpotent groups, _Discrete Math._ 54 (1985), 15-21.

P. Lorimer, Vertex-transitive graphs of valency 3, _Europ. J. Combinatorics_ (1983), 37-44.

———, Vertex-transitive graphs : symmetric graphs of prime valency, _J. Graph Theory_ 8, no.1 (1984), 55-68.

———, Trivalent symmetric graphs of order at most 120, _Europ. J. Combinatorics_ 5 (1984), 163-171.

L. Lovasz, Problem 11, _Combinatorial Structures and Their Applications_, Gordon and Breach, New York, 1970.

———, _Combinatorial Problems and Exercises_, Akadémiai Kiado, Budapest, 1979.

W. Mader, Über den Zusammenhang symmetrischer Graphen, _Arch. Math._ (Basel) 21 (1970), 331-336.

D. Marušič, Hamiltonian circuits in Cayley graphs, _Discrete Math._ 46 (1983), 49-54.

D. Marušič and T. D. Parsons, Hamiltonian paths in vertex-symmetric graphs of order 5p, _Discrete Math._ 42 (1981), 227-242.

———, Hamiltonian paths in vertex-symmetric graphs of order 4p, _Discrete Math._ 43 (1983), 91-96.

M. H. McAndrew, On graphs with transitive automorphism groups, _Notices Amer. Math. Soc._ 12 (1965), 575.

B. D. McKay, Transitive graphs with fewer than twenty vertices, _Math. of Computation_ 33 (1979), 1101-1121.

E. Mendelsohn, Every (finite) group is the group of automorphism of a (finite) strongly regular graph, _Ars Combin._ 6 (1978), 75-86.

G. H. T. Meredith, Triple transitive graphs, _J. London Math. Soc._, Ser.2, 13 (1976), 249-257.

R. C. Miller, The trivalent symmetric graphs of girth at most six, J. Combin. Theory 10 (1971), 163-182.

S. P. Mohanty, M. R. Sridharan and S. K. Shukla, On cycle permutation groups and graphs, J. Math. Phy. Sci. 12 (1978), 409-416.

A. Mowshowitz, The group of a graph whose adjacency matrix has distinct eigenvalues, Proof Techniques in Graph Theory (Ed. F. Harary), Academic Press, New York (1969), 109-110.

D. S. Passman, Permutation Groups, W. A. Benjamin, New York, 1968.

M. Petersdorf and H. Sachs, Spektrum und Automorphismengruppe eines Graphen, Combinatorial Theory and its Applications III, North-Holland (1970), 891-907.

C. E. Praeger, When are symmetric graphs characterised by their local properties? Combinatorial Mathematics IX (Eds. E. J. Billington, S. Oates-Williams and A. P. Street), Springer-Verlag Lecture Notes in Math. 952 (1982), 123-141.

———, Graphs and their automorphism groups, Algebraic Structures and Applications, Lecture Notes in Pure and Appl. Math. 74, Dekker, New York (1982), 21-32.

L. V. Quintas, Extrema concerning asymmetric graphs, J. Combin. Theory 3 (1967), 57-82.

R. J. Riddell, Contributions to the theory of condensation, Dissertation, Univ. of Michigan, Ann Arbor, 1951.

C. Ronse, On homogeneous graphs, J. London Math. Soc., Ser.2, 17 (1978), 375-379.

G. Sabidussi, Graphs with given group and given graph theorectical properties, Canad. J. Math. 9 (1957), 515-525.

———, On a class of fix-point-free graphs, Proc. Amer. Math. Soc. 9 (1958), 800-804.

———, On the minimum order of graphs with given automorphism group, Monatsh. Math. 63 (1959), 124-127.

———, Graphs with given infinite group, Monatsh. Math. 64 (1960), 64-67.

———, Vertex-transitive graphs, Monatsh. Math. 68 (1964), 426-438.

S. C. Shee and H. H. Teh, H-extension of graphs, Combinatorica 4 (1984), 207-211.

J. Sheehan, Smoothly embeddable subgraphs, J. London Math. Soc., Ser.2, 9 (1974), 212-218.

S. Shelah, Graphs with prescribed asymmetry and minimal number of edges, Infinite and finite sets, Colloq. Math. Soc. Jànos Bolyai 10, (1975), 1241-1256.

C. C. Sims, Graphs and finite permutation groups, Math. Zeistschr. 95 (1967), 76-86.

————, ibid. II, 103 (1968), 276-281.

B. P. Sinka and P. K. Srimani, Some studies on the characteristic polynomial of a graph, Internat. J. Electron. 54 (1983), 377-400.

D. H. Smith, Distance-transitive graphs of valency four, J. London Math. Soc. 8 (1974), 377-384.

————, Bounding the diameter of a distance-transitive graph, J. Combin. Theory 16 (1974), 139-144.

M. S. Smith, On rank 3 permutation groups, J. Algebra 33 (1975), 22-42.

————, A characterization of a family of rank 3 permutation groups, Quart. J. Math. Oxford Ser.(2) 26 (1975), 99-105.

K. H. Tan and H. H. Teh, Finite Auto-extensions of graphs, Nanta Math. vol.3, no.2 (1969), 33-42.

H. H. Teh, A fundamental theorem concerning the chromatic number of a group graph, Nanta Math., vol.1 (1966-67), 76-81.

————, An algebraic representation theorem of finite regular graphs, Nanta Math., vol.3, no.2 (1969), 88-94.

H. H. Teh and S. L. Ang, Finite homogeneous auto-extensions of graphs, Nanta Math. 1 (1966/67), 82-86.

H. H. Teh and C. C. Chen, Point symmetric ranks of undirected graphs, Nanta Math. 4 (1970), 85-100.

H. H. Teh and S. C. Shee, Algebraic Theory of Graphs, Lee Kong Chian Institute of Mathematics and Computer Science, Nanyang Univ., Singapore, 1976.

H. H. Teh and H. P. Yap, Some construction problems of homogeneous graphs, Bull. Math. Soc., Nanyang Univ., (1964), 164-196.

J. Turner, Point-symmetric graphs with a prime number of points, J. Combin. Theory 3 (1967), 136-145.

W. T. Tutte, A family of cubical graphs, Proc. Camb. Phil. Soc. 43 (1947), 459-474.

————, The Connectivity of Graphs, Toronto University Press, Toronto, 1966.

K. S. Vijayan, On a class of distance transitive graphs, J. Combin. Theory, Ser.B, 25 (1978), 125-129.

————, On arc-transitive graphs, Combinatorics, Keszthely 1976, Colloq. Math. Janos Bolyai 18 (1978), 1091-1098.

W. Vogler, Graphs wtih given group and given constant link, J. Graph Theory 9 (1984), 111-115.

M. E. Watkins, Some classes of hypoconnected vertex-transitive graphs, Recent Progress in Combinatorics (Eds., W. T. Tutte and C. St. J. A. Nash-Williams), Academic Press, New York (1969), 323-328.

————, Connectivity of transitive graphs, J. Combin. Theory 8 (1970), 23-29.

————, On the action of non-abelian groups on graphs J. Combin. Theory, Ser.B, 11 (1971), 95-104.

R. M. Weiss, Über s-reguläre Graphen, J. Combin. Theory, Ser.B, 16 (1973), 229-233.

————, Über lokal s-reguläre Graphen, J. Combin. Theory, Ser.B, 20 (1976), 124-127.

————, Glatt einbettbare Untergraphen, J. Combin. Theory, Ser.B, 21 (1976), 275-281.

————, Über symmetrische Graphen und die projektiven Gruppen, Arch. Math. (Basel) 28, no.1 (1977), 110-112.

————, s-transitive graphs, Algebraic Methods in Graph Theory, Colloq. Math. Soc. Jànos Bolyai, Szeged, Hungary (1978), 827-847.

————, The nonexistence of 8-transitive graphs, Combinatorica 1, no.3 (1981), 309-311.

————, Edge-symmetric graphs, J. Algebra 75 (1982), 261-274.

H. Whitney, Congruent graphs and the connectivity of graphs Amer. J. Math. 54 (1932), 150-168.

D. Witte, On hamiltonian circuits in Cayley diagrams, Discrete Math. 38 (1982), 99-108.

W. J. Wong, Determination of a class of primitive permutation groups, Math. Zeitschr. 99 (1967), 235-246.

R. W. Woodrow, There are four countable ultrahomogeneous graphs without triangles, J. Combin. Theory, Ser.B, 27 (1979), 168-179.

E. M. Wright, Graphs on unlabelled nodes with a given number of edges, Acta Math. 126 (1971), 1-9.

———, Asymmetric and symmetric graphs, Glasgow Math. J. 15 (1974), 69-73.

H. P. Yap, Some addition theorems of group theory with applications to graph theory, Canad. J. Math. 22 (1970), 1185-1195.

———, Point-symmetric graphs with $p < 13$ points, Nanta Math. vol.6, no.1 (1973), 8-20.

———, The characteristic polynomial of the adjacency matrix of a multi-digraph, Nanta Math., vol.8, no.1 (1975), 41-46.

———, On 4-ultratransitive graphs, Soochow J. Math. 9 (1983), 97-109.

4. PACKING OF GRAPHS

1. Introduction and definitions

Suppose $G_1, G_2, \ldots, G_k$ are graphs of order at most n. We say that there is a packing of $G_1, G_2, \ldots, G_k$ into the complete graph K_n if there exist injections $\alpha_i : V(G_i) \to V(K_n)$, $i = 1,2,\ldots, k$ such that $\alpha_i^*(E(G_i)) \cap \alpha_j^*(E(G_j)) = \phi$ for $i \neq j$, where the map $\alpha_i^* : E(G_i) \to E(K_n)$ is induced by α_i. Similarly, suppose G is a graph of order m and H is a graph of order $n \geq m$ and there exists an injection $\alpha : V(G) \to V(H)$ such that $\alpha^*(E(G)) \cap E(H) = \phi$. Then we say that there is a packing of G into H, and in case n = m, we also say that there is a packing of G and H or G and H are packable. Thus G can be packed into H if and only if G is embeddable in the complement $\bar{H}$ of H. However, there is a slight difference between embedding and packing. In the study of embedding of a graph into another graph, usually at least one of the graphs is fixed whereas in the study of packing of two graphs very often both the graphs are arbitrarily chosen from certain classes.

In practice, one would like to find an efficient algorithm to pack two graphs G and H. But this has been shown to be an NP-hard problem (see Garey and Johnson [79;p.64]). However, if one restricts G and H to some special classes of graphs, for instance, both G and H are trees, then there exist polynomial time algorithms for the packing of G and H (mentioned in Hedetniemi, Hedetniemi and Slater [81]).

In this chapter, we shall present some results on the following two theoretical packing problems. The first one is a dense packing of trees of different sizes into K_n (§2). The second one is a general packing of two graphs having small size (§3 to §8). The main reference of this chapter is the last chapter of Bollobàs' book : "Extremal Graph Theory". Some results of this chapter will be applied in Chapter Five.

We shall require the following definition. Suppose there is a packing α of G and H. Then the set of edges of H incident with at least one vertex of $\alpha(v)$, $v \in V(G)$, is said to be covered by G.

Exercise 4.1

1. Let T_i be a tree of order i. Prove that any sequence of trees T_2, T_3, ..., T_n, in which all but two are stars, can be packed into K_n (Gyárfás and Lehel [78]).

2. Let G be a graph of order n and size n - 2. Prove that there exists a packing α of G into its complement $\bar{G}$ such that $\alpha(v) \neq v$ for all $v \in V(G)$ (Burns and Schuster [77]).

3.* **Ringel's conjecture** For any arbitrary tree T_{n+1} of order n+1, there is a packing of 2n + 1 copies of T_{n+1} into K_{2n+1} (Ringel [63]).

2. Packing n - 1 trees of different size into K_n

Erdös and Gallai [59] proved that every graph G having size $e(G) > \frac{1}{2}|G|(k - 1)$ contains a path of length k. In 1963, Erdös and Sós made the following conjecture.

Conjecture (Erdös and Sós) If G is a graph and $e(G) > \frac{1}{2}|G|(k - 1)$, then G contains every tree of size k.

Motivated by the above conjecture, Gyárfás raised the question that whether any sequence of trees T_2, ..., T_n (T_i is a tree of order i) can be packed into K_n. Gyárfás' question has now been quoted in many books and research papers as the Tree Packing Conjecture (TPC).

Tree Packing Conjecture Any sequence of trees T_2, T_3, ..., T_n can be packed into K_n.

Gyárfás and Lehel [78] proved that the TPC is true for two special cases. The first special case is given in Ex.4.1(1). Their proof of the second special case (Theorem 2.1) is complicated. An alternate, short proof of the second special case due to Zaks and Liu [77] is given below. Straight [79] also confirmed that the TPC is true for several other special cases. His results are given in Ex.4.2(5), (6) and (7). From straight's results, we can verify that the TPC is true for all $n \leq 7$. Fishburn [83] also proved that the TPC is true for some more special cases and verified that the TPC is true for n = 8, 9. We note that if the TPC fails to hold for some integer N, then it fails to hold for all integers $n > N$. (For $n = N + 1$, simply take $T_n = S_n$.)

We need the following definitions. Let $A[G] = (a_{ij})$ be the adjacency matrix of a graph G. Since G is undirected, A[G] is symmetric and thus we need only to use the upper right part of A[G] which is denoted by $A^R[G]$. In $A^R[G]$ we name the following sequence of 1's as a (right) <u>stair</u> :

$$a_{i,j},\ a_{i,j+1},\ a_{i-1,j+1},\ a_{i-1,j+2},\ \cdots,\ a_{i-k,j+k},\ a_{i-k,j+k+1}$$

where it is possible that either $a_{i,j}$ or $a_{i-k,j+k+1}$ (or both) may be excluded. From this definition, it follows that a stair in $A^R[G]$ corresponds to a path in G.

Theorem 2.1 <u>Any sequence of trees</u> $T_2, \ldots, T_n$ <u>where</u> $T_i \in \{S_i, P_i\}$ <u>can be packed into</u> K_n.

Proof. We partition $A^R[K_n]$ into the following two blocks :

$A_1(n) = \{a_{ij} \mid i + j < n\}$, $A_2(n) = \{a_{ij} \mid i + j > n\}$ (see Fig.4.1).

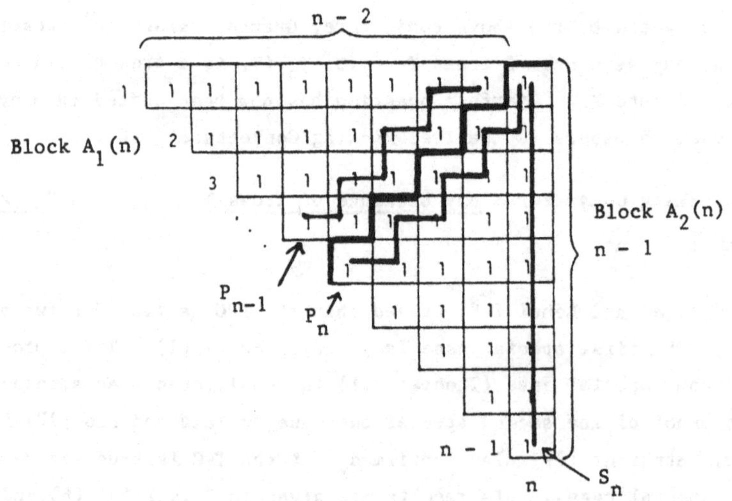

Figure 4.1

We prove the theorem for even n (for odd n, the proof is similar). We shall show that $A_1(n)$ can be decomposed corresponding to the given trees T_i for odd i, and that $A_2(n)$ can be decomposed corresponding to the trees T_j for even j. The proofs for $A_1(n)$ and $A_2(n)$ are similar, so let us show the result for $A_2(n)$ by induction on n.

For n = 2, it is clear that $T_2 = S_2$ can be packed into K_2. Assume that the result is true for all even $m < n$. Now by deleting either the last column (which corresponds to the case $T_n = S_n$) or the stair in $A_2(n)$ as shown in Fig.4.1 (which corresponds to the case $T_n = P_n$), and by deleting the first row (which corresponds to the case $T_{n-1} = S_{n-1}$) or the stair in $A_1(n)$ as shown in Fig.4.1 (which corresponds to the case $T_{n-1} = P_{n-1}$), we are left with two blocks of $A_1(n-2)$ and $A_2(n-2)$. The result follows by induction. //

Since a resolution to the TPC is hard, we may look for some other aspects of a similar nature. Theorem 2.3 is an example of such an attempt. We shall require the following lemma (see Bollobás [78; p.xvii]).

Lemma 2.2 *Let* k *be a positive integer. Suppose* H *is a graph of order* $n \geq k + 1$. *If*

$$e(H) \geq (k - 1)(n - k) + \binom{k}{2} + 1 ,$$

then H *contains a subgraph* F *such that* $\delta(F) \geq k$.

Proof. We prove this lemma by induction on n. Suppose $n = k + 1$. Then

$$\binom{k+1}{2} \geq e(H) \geq (k - 1) + \binom{k}{2} + 1 = \binom{k+1}{2}.$$

Hence $H = K_{k+1}$ and the result follows.

Suppose $n \geq k + 2$ and $\delta(H) \leq k - 1$. Let x be a vertex of H. Then

$$\begin{aligned} e(H - x) &\geq (k - 1)(n - k) + \binom{k}{2} + 1 - (k - 1) \\ &= (k - 1)[(n - 1) - k] + \binom{k}{2} + 1, \end{aligned}$$

and by induction, $H - x$ contains a subgraph F such that $\delta(F) \geq k$. //

Theorem 2.3 (Bollobás [83]) If $3 < s < \frac{1}{2}\sqrt{2}\, n$, then any sequence of trees $T_2, T_3, \ldots, T_s$ can be packed into K_n.

Proof. Suppose $T_{k+1}, T_{k+2}, \ldots, T_s$ have been packed into K_n. Let $H = K_n - \cup_{j=k+1}^{s} E(T_j)$. Then H is a graph of order n and

$$e(H) = \frac{1}{2}\{n^2 - n - (s + k - 1)(s - k)\} \qquad (1)$$

We claim that H has a subgraph F having minimum valency k - 1. Indeed, otherwise by Lemma 2.2, we have

$$e(H) < \binom{k-1}{2} + (k - 2)(n - k + 1) \qquad (2)$$

Now (1) and (2) imply that

$$n^2 - s^2 - 2kn + 2k^2 + 3n + s + 2 - 4k < 0 \qquad (3)$$

Since $s^2 < \frac{1}{2}n^2$, from (3) we have

$$n^2 - 4kn + 4k^2 + 6n + 2s + 4 - 8k < 0,$$

from which it follows that

$$(n - 2k)^2 + 6n + 2s + 4 - 8k < 0,$$

which is false because $n > s > k + 1$.

Finally, $\delta(F) > k - 1$ implies that T_k can be embedded in F. //

Remark. If the conjecture of Erdös and Sós is true, then we can replace the bound $\frac{1}{2}\sqrt{2}\, n$ by $\frac{1}{2}\sqrt{3}\, n$.

Exercise 4.2

1. Prove that the sequence of paths $P_2, P_4, \ldots, P_{2n}$ can be packed into $K_{n,n}$ (Zaks and Liu [77]; Fink and Straight [81]).
2. Prove that for odd n, the sequence of paths $P_3, P_5, \ldots, P_{2n+1}$ can be packed into $K_{n,n+1}$ (Zaks and Liu [77]; Fink and Straight [81]).
3. Prove that any sequence of trees $T_2, T_3, \ldots, T_n$ where $T_i \in \{S_i, P_i\}$

can be packed into $K_{\frac{n}{2},n-1}$ ($K_{\frac{n-1}{2},n}$) for n even (odd) (Zaks and Liu [77]).

4. Prove that any sequence of trees T_2, T_3, ..., T_n such that each T_i having diameter at most three can be packed into K_n (Hobbs [81]). (A. M. Hobbs and J. Kasiraj have improved this result by allowing one of the trees T_i to have diameter more than three - mentioned in Hobbs [81].)

5. Let L_i be the tree of order i shown in Fig.4.2.

Figure 4.2

Prove that any sequence of trees $T_2, \ldots, T_n$ where $T_i \in \{S_i, P_i, L_i\}$ can be packed into K_n (Straight [79]).

6. Prove that for any integer $n \geq 2$, any sequence of trees $T_2, \ldots, T_n$ such that $\Delta(T_i) \geq i - 2$ with at most two exceptions, can be packed into K_n (Straight [79]).

7. Prove that for any integer $n \geq 2$, any sequence of trees $T_2, \ldots, T_n$ such that $\Delta(T_i) \geq i - 3$ with at most one exception, can be packed into K_n (Straight [79]).

8. Using the results of problems 5, 6 and 7, prove that the TPC is true for all $n \leq 7$ (Straight [79]).

9.$^+$ A vertex of a tree having valency at least two is an <u>interior vertex.</u> A tree is a <u>caterpillar</u> if the induced subgraph of its interior vertices is a path. A diameter-3 (diameter-4) caterpillar is called a <u>double star</u> (<u>triple star</u>). A triple star is <u>unimodal</u> if its interior vertices x, y and z satisfy $d(y) \geq d(x)$ or $d(y) \geq d(z)$ where xyz is a path. An <u>interior-3 caterpillar</u> is a caterpillar having all interior vertices x_1, x_2, ..., x_n such that $x_1 x_2 \ldots x_n$ is a path and $d(x_1) \geq 2$, $d(x_2) = \ldots = d(x_{n-1}) = 3$ and $d(x_n) = 2$. A tree is a <u>spider</u> if it has only one vertex of valency at least three. A <u>thin-body spider</u> is a spider such that one and only one of its end-vertices has diameter at least 3 from the body vertex (see Fig.4.3).

Prove that if each T_i is a double star, unimodal triple star, interior-3 caterpillar or thin-body spider, then $T_2, \ldots, T_n$ can be packed into K_n (Fishburn [83]).

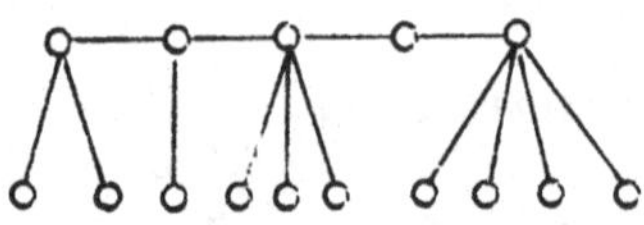

a caterpillar

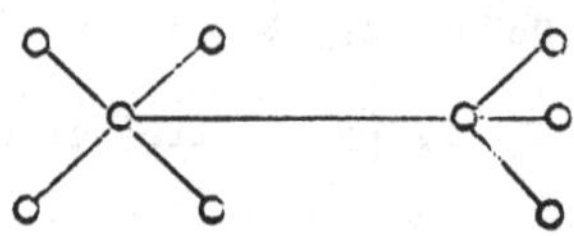

a double star

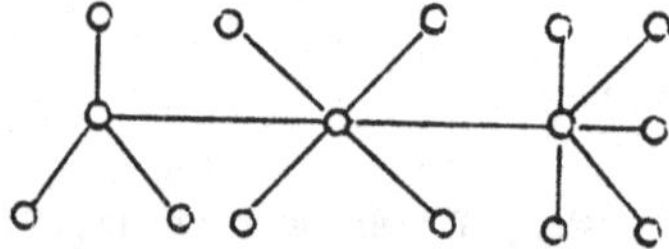

a unimodal triple star

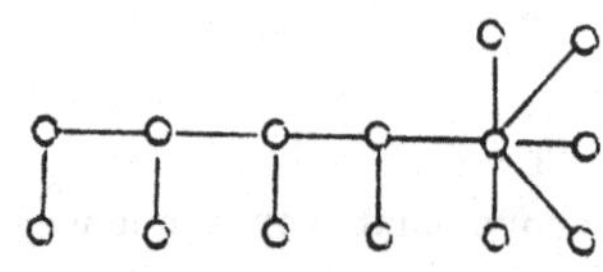

an interior-3 caterpillar

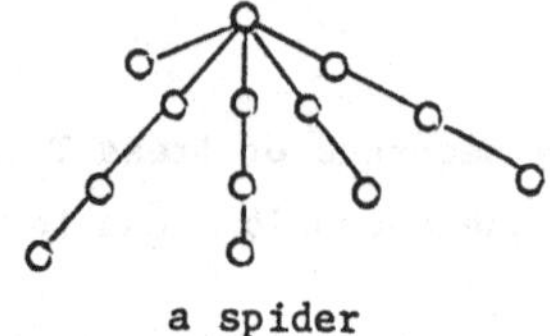

a spider

a thin-body spider

Figure 4.3

10. Using the result of Problem 9, prove that the TPC is true for $n = 8$ and $n = 9$ (Fishburn [83]).

11. Let $k \geq 3$ and $n \geq k + 1$ be integers. Suppose G is a graph of order n and $e(G) \geq \frac{1}{2}(1 + \frac{n-k}{n})n(k - 1) + 1$. Prove that G contains all trees of size k. Hence, show that if k and n are integers satisfying $\frac{n}{2} \leq k < \frac{2}{3}n$, and G is a graph of order n and size at least $\frac{3}{4}n(k - 1)$, then G contains all trees of size k.

3. Packing two graphs of small size

In this section we present some sufficient conditions for packing two graphs of small size. We shall apply these results in the subsequent sections.

Theorem 3.1 (Sauer and Spencer [78]) *Suppose* G *and* H *are graphs of order* n. *If* $e(G)e(H) < \binom{n}{2}$, *then* G *and* H *are packable*.

Proof. Let Σ be the set of all bijections $\alpha : V(G) \to V(H)$. Then $|\Sigma| = n!$. Let $e = \{u,v\} \in E(G)$ and $\alpha \in \Sigma$. The induced map α^* of α on $E(G)$ is given by $\alpha^*(e) = \{\alpha(u),\alpha(v)\}$. For any $e \in E(G)$ and $f \in E(H)$, we define

$$A(e,f) = \{\alpha \in \Sigma \mid \alpha^*(e) = f\}.$$

Then $\left|\bigcup_{e,f} A(e,f)\right| \leq \sum_{e,f} |A(e,f)| = 2e(G)e(H)(n-2)! < n!$.

Hence there exists $\sigma \in \Sigma$ such that $\alpha^*(e) \neq h$ for any $e \in E(G)$ and for any $f \in E(H)$, i.e. σ is a packing of G and H. //

Remark. Let $G = S_{2m}$ and $H = mK_2$. Then G cannot be packed in H. Thus Theorem 3.1 is best possible.

The following theorem was first announced by Catlin [74]. The proof given here is due to Sauer and Spencer [78] (see also Bollobás [78;p.425]).

Theorem 3.2 *Suppose* G *and* H *are graphs of order* n. *If* $2\Delta(G)\Delta(H) < n$, *then there is a packing of* G *and* H.

Proof. The assertion is trivial for $n \leq 2$ so we assume that $n \geq 3$. Let $\alpha : V(G) \to V(H)$ be a bijection such that G and H have minimum number of edges in common. Suppose $V(G) = \{x_1, \ldots, x_n\}$ and $V(H) = \{y_1, \ldots, y_n\}$ where $y_i = \alpha(x_i)$. We may assume that $x_1x_2 \in E(G)$ and $y_1y_2 \in E(H)$ and obtain a contradiction by showing that there exists an index $i > 2$ such that by flipping x_2 with x_i, i.e. mapping x_2 to y_i and x_i to y_2, the number of common edges decreases.

Let L be the set of indices $k > 2$ for which there is an index j such that either $x_2x_j \in E(G)$ and $y_ky_j \in E(H)$ or $x_kx_j \in E(G)$ and $y_2y_j \in E(H)$. Since $d(x_2), d(x_k) \leq \Delta(G)$, $d(y_k), d(y_2) \leq \Delta(H)$, $x_1x_2 \in E(G)$ and $y_1y_2 \in E(H)$, $|L| \leq 2\Delta(G)\Delta(H) - 2 < n - 2$. Hence there exists an index i, $3 \leq i \leq n$, such that $i \notin L$. We can now flip x_2 with x_i. After flipping, there are no edges of G incident with x_2 or x_i overlapping with any edge of H. The other edges of G and H remain the same. Thus the number of overlapping edges of G and H have been decreased, a contradiction. //

Bollobás and Eldridge [78a] pointed out that the result of Theorem 3.2 is almost best possible. For example, suppose d_1 and d_2 are two integers such that $d_1 \leq d_2 < n$ and suppose $n \leq (d_1 + 1)(d_2 + 1) - 2$. Let $n = p_i(d_i + 1) + r_i$, $1 \leq r_i \leq d_i + 1$, $G = p_1K_{d_1+1} \cup K_{r_1}$, and $H = p_2K_{d_2+1} \cup K_{r_2}$. Then $\Delta(G) = d_1$, $\Delta(H) = d_2$ and there is no packing of G and H. Thus if $n \leq (\Delta(G) + 1)(\Delta(H) + 1) - 2$, then G and H may not be packable. Motivated by this example, they made the following conjecture.

Conjecture (Bollobás and Eldridge) *If* G *and* H *are graphs of order* n *and* $(\Delta(G) + 1)(\Delta(H) + 1) \leq n + 1$, *then there is a packing of* G *and* H.

Remarks.

(1) From Theorem 3.2 it follows that this conjecture is true for $\Delta(G) = 1$.

(2) It is mentioned in Bollobás [78; p.426] that the trueness of this conjecture for $\Delta(G) = 2$ implies a theorem of Corrádi and Hajnal; and the trueness of this conjecture extends a theorem of Hajnal and Szemerédi. For detail, see Catlin [77].

(3) For further conjectures and open problems on the packing of two graphs, the readers may refer to Bollobás [78; p.436 - 437].

Exercise 4.3

1. Let m and s be positive integers such that

$$\binom{s}{2} \leq m < \binom{s+1}{2} .$$

Let G and H be graphs of order n, e(G) = m and

$$e(H) \leq \binom{n}{2} - t_{s-1}(n) - 1$$

where $t_{s-1}(n)$ is the size of the Turan graph $T_s(n)$ which is approximately equal to $\frac{1}{2}(1 - \frac{1}{s-1})n^2$. Prove that if n is sufficiently large, then there is a packing of G and H (Bollobás and Eldridge [78a]. See also Bollobás [78; p.426]).
(Note that the example $G = K_s \cup O_{n-s}$ and $H = \bar{T}_{s-1}(n)$ shows that this result is best possible.)

2. Suppose $0 < \alpha < \frac{1}{2}$ and n is sufficiently large. Prove that if G and H are graphs of order n, $e(G) \leq \alpha n$ and $e(H) \leq \frac{1}{5}(1 - 2\alpha)n^{3/2}$, then there is a packing of G and H (Bollobás and Eldridge [78a]. See also Bollobás [78; p.427]).
(Note that the example $G = K_{t+1} \cup O_{n-t-1}$ and $H = tK_{n/t}$ shows that this result is near to being best possible.)

4. Packing two graphs of order n having total size at most 2n − 3

Milner and Welsh [74] noticed that if any two graphs G and H of order n such that $e(G) + e(H) \leq [\frac{3}{2}(n-1)]$ are packable, then one can prove that the computational complexity of any graph property F has lower bound $[\frac{3}{2}(n-1)]$. They conjectured that $e(G) + e(H) \leq [\frac{3}{2}(n-1)]$ is sufficient for the packing of two graphs G and H. This conjecture was proved by Sauer and Spencer [78]. We now present the main packing theorem of Bollobás and Eldridge [78a] which generalizes the result of Sauer and Spencer. The proof of the main packing theorem depends heavily on Lemma 4.1. The alternate proof of Lemma 4.1 and a slight simplification of the original proof of Theorem 4.2 given here are due to Teo [85].

Lemma 4.1 Let T be a tree of order p and let G be a graph of order n. Suppose $2 \leq 2p \leq n$ $(n \geq 5)$, $\Delta(G) < n - 1$ and $n - 1 \leq e(G) \leq n + \frac{n}{p} - 3$. Then there is a packing of T and G such that T covers at least p + 1 edges of G and $\Delta(G - T) < n - p - 1$.

Proof. We shall prove this lemma by induction on n and p. Suppose

$p = 1$ and T is the isolated vertex x. We can map x to a vertex u of G such that $d(u) = \Delta(G)$. If $\Delta(G - x) < n - 2$, we are done. Otherwise G has two vertices u and v such that $d(u) = d(v) = n - 2$ and $N(u) \cap N(v) \neq \phi$. We now map x to w, where $w \in N(u) \cap N(v)$.

Suppose $p = 2$ and T is the edge xy. We now map x to a vertex u of valency $\Delta(G)$ and map y to a vertex v not adjacent to u and of maximum possible valency. Then T covers at least three edges of G. If $G - T$ has a vertex w of valency $n - 3$, then u is adjacent to w for otherwise both u and v have valency at least $n - 3$ and so $3(n - 3) \leq e(G) \leq \frac{3}{2}(n - 2)$, i.e. $n \leq 4$, a contradiction to the assumption that $n \geq 5$. Consequently both u and w have valency $n - 2$ and so $2(n - 2) - 1 \leq \frac{3}{2}(n - 2)$, i.e. $n \leq 4$, another contradiction. Hence $\Delta(G - T) < n - 3$.

Suppose $p \geq 3$. Then $n \geq 6$. Let x be an end-vertex of T, y be the neighbour of x, $T' = T - x$ and $T'' = T - x - y$. Since $e(G) \leq n + \frac{n}{3} - 3$, $\delta(G) \leq 2$. We consider the following three cases separately.

Case 1. $\delta(G) = 0$.

Let u and v be vertices of G such that $d(u) = 0$ and $d(v) = \Delta(G)$. Clearly $\Delta(G - u - v) < n - 3$ for otherwise $2(n - 3) \leq e(G) \leq n + \frac{n}{3} - 3$ implies that $n \leq \frac{9}{2}$. By adding $\Delta(G) - 2$ edges to $G - u - v$ in a nice way, we obtain a graph G' such that $\Delta(G') < n - 3$ and $n - 3 \leq e(G') \leq n - 2 + \frac{n-2}{p-2} - 3$. By induction (the case that $n = 6$ can be verified easily), there is a required packing σ of T'' and G'. We can extend σ to a required packing of T and G by letting $\sigma(x) = v$ and $\sigma(y) = u$.

Case 2. $\delta(G) = 1$.

Let u and v be vertices of G such that $d(u) = 1$ and $uv \in G$. Suppose $\Delta(G - u) = n - 2$. Then G has a vertex $z \neq v$ such that $d(z) = n - 2$. Now

$$e(G - u - v - z) \leq n + \frac{n}{p} - 3 - (n - 1) = \frac{n}{p} - 2 \quad \text{and}$$

$$e(T'')e(G - u - v - z) \leq (p - 3)(\frac{n}{p} - 2) \leq n - 6 < \binom{n-3}{2}.$$

Thus by Theorem 3.1, there is a packing σ of T'' and $G - u - v - z$. We can extend σ to a required packing of T and G by letting $\sigma(y) = u$ and

$\sigma(x) = z$. (Note that T covers at least $n - 1 \geq \frac{n}{2} + 1 \geq p + 1$ edges and $\Delta(G - T) \leq \frac{n}{p} - 2 < \frac{n}{2} - 1 \leq n - p - 1$.)

Suppose $\Delta(G - u) < n - 2$. We have $(n - 1) - 1 \leq e(G - u) \leq n - 1 + \frac{n}{p} - 3 < n - 1 + \frac{n-1}{p-1} - 3$. If T is a tree of order $s \leq p - 1 \leq \frac{n-1}{2}$, then by induction, there is a required packing of T and G - u which is also a required packing of T and G. Hence we need only to consider the case $|T| = p = \frac{n}{2}$. In this case $e(G) = n - 1$. Now if $d(v) < \frac{n}{2}$, then since there is a required packing σ of T' and G - u, we can extend σ to a required packing of T and G by letting $\sigma(x) = u$ (if $\sigma(y) \neq v$) or by mapping x to a vertex not adjacent to v (if $\sigma(y) = v$). Finally, if G has no such pair of vertices u and v, then since $e(G) = n - 1$, $G = S_k \cup C_{r_1} \cup \dots \cup C_{r_j}$ where $k \geq \frac{n}{2} + 1$. It is easy to obtain a required packing of T and G by mapping a vertex of T to a vertex of a cycle-component of G and the other vertices of T to the vertices of the star-component.

Case 3. $\delta(G) = 2$.

Let u be a vertex of G having valency 2. The case $\Delta(G - u) = n - 2$ can be settled as in Case 2. Hence we assume that $\Delta(G - u) < n - 2$. Since $\delta(G) = 2$, $e(G) \geq n$. Hence $(n - 1) - 1 \leq e(G - u) \leq n - 2 + \frac{n}{p} - 3 < n - 1 + \frac{n-1}{p} - 3$. Thus, by induction, there exists a required packing of T and G - u unless $2p = n$. However, if $2p = n$, then $e(G) = n - 1$, a contradiction. //

Theorem 4.2 _Suppose_ H _and_ G _are graphs of order_ n, $\Delta(H), \Delta(G) < n - 1$, $e(H) + e(G) \leq 2n - 3$ _and_ $\{H,G\}$ _is not one of the following pairs_ :

$\{2K_2, O_1 \cup K_3\}$, $\{O_2 \cup K_3, K_2 \cup K_3\}$, $\{3K_2, O_2 \cup K_4\}$, $\{O_3 \cup K_3, 2K_3\}$, $\{2K_2 \cup K_3, O_3 \cup K_4\}$, $\{O_4 \cup K_4, K_2 \cup 2K_3\}$ _and_ $\{O_5 \cup K_4, 3K_3\}$.

Then there is a packing of H _and_ G.

Proof. Without loss of generality we assume that $e(H) + e(G) = 2n - 3$, $e(H) \leq n - 2$ and $e(G) \geq n - 1$. Let p be the order of the smallest tree-component T (which may be O_1) of H. Then $2p \leq n$ and $e(H) \geq n - \frac{n}{p}$. Hence $e(G) \leq n + \frac{n}{p} - 3$. Now if $n \geq 5$, then by Lemma 4.1, there is a packing of T and G such that $e(H - T) + e(G - T) \le 2(n - p) - 3$ and $\Delta(G - T) < n - p - 1$. Thus, if $\Delta(H - T) < n - p - 1$ and $\{H - T, G - T\}$ is not one of the forbidden pairs, then by induction $H - T$ can be packed into $G - T$ and so H can be packed into G. It remains to consider three special cases, namely, (i) $n \le 4$; (ii) $\Delta(H - T) \ge n - p - 1$; and (iii) $\{H - T, G - T\}$ is one of the forbidden pairs.

Suppose $n \le 4$. Then $p \le 2$. The case $p = 1$ is trivial. On the other hand if $p = 2$, then $n = 4$, $e(H) = 2$ (and thus $H = 2K_2$) and $e(G) = 3$. In this case H and G are packable unless $G = O_1 \cup K_3$.

Suppose $\Delta(H - T) \ge n - p - 1$. Then $n - 2 - (p - 1) \ge e(H - T) \ge n - p - 1$, from which it follows that $e(H - T) = n - p - 1$ and thus $H - T = S_{n-p}$. From this, we have $e(H) = n - 2$ and $e(G) = n - 1$. We can now map the centre of the star $H - T$ to a vertex u of G such that $d_G(u) \le 1$ and map the rest of the vertices arbitrarily so that they cover at least $n - p + 1$ edges of G and $\delta(G - (H - T)) = 0$. Hence $e(T) + e(G - (H - T)) \le (p - 1) + (n - 1 - (n - p + 1)) = 2p - 3$ and $\Delta(G - (H - T)) \le e(G - (H - T)) < p - 1$. If T is a star, we can pack the centre of T on an isolated vertex of $G - (H - T)$. Otherwise $\Delta(T) < p - 1$ and by induction we can pack T with $G - (H - T)$.

Finally, suppose $\{H - T, G - T\}$ is one of the forbidden pairs. We first observe that $e(H - T) \le (n - 2) - (p - 1) = n - p - 1$. We shall begin with the smallest n to obtain the other forbidden pairs in a systematic way as indicated below (we prove only for two cases, the other cases can be proved in a similar way).

Suppose $\{H - T, G - T\} = \{2K_2, O_1 \cup K_3\}$. If $H - T = O_1 \cup K_3$, then by the choice of T, we have $T = O_1$. Hence $H = O_2 \cup K_3$ and G is a subgraph of $O_1 + (2K_2)$ with $e(G) = 4$. Thus H and G are packable unless $G = K_2 \cup K_3$. On the other hand, if $H - T = 2K_2$, then either $H = O_1 \cup 2K_2$ or $H = 3K_2$. Suppose $H = O_1 \cup 2K_2$. Then G is a subgraph of $O_1 + (O_1 \cup K_3)$ with $e(G) = 5$. Thus H and G are packable. Suppose $H = 3K_2$. Then G is a subgraph of $O_2 + (O_1 \cup K_3)$ with $e(G) = 6$. Thus H and G are packable unless $G = O_2 \cup K_4$.

Suppose $\{H - T, G - T\} = \{O_5 \cup K_4, 3K_3\}$. Then $H - T = O_5 \cup K_4$ because $e(H - T) \le n - p - 1$. Hence $H = O_6 \cup K_4$ and G is subgraph of $O_1 + (3K_3)$ with $e(G) = 11$. Thus H and G are packable. //

Suppose G and H are graphs of order n. If $\Delta(G) = n - 1$ and $\delta(H) \ge 1$ (or $\Delta(H) = n - 1$ and $\delta(G) \ge 1$), then obviously there is no packing of G and H. Hence in Theorem 4.2 we have to assume that $\Delta(H), \Delta(G) < n - 1$. Now from Theorem 4.2 we have

Corollary 4.3 *Suppose* G *and* H *are graphs of order* n *such that neither* $\Delta(H) = n - 1$ *and* $\delta(G) \ge 1$ *nor* $\Delta(H) = n - 1$ *and* $\delta(G) \ge 1$ *holds. We have*

(i) *if* $e(G) + e(H) \le 2n - 3$ *and* {G, H} *is not one of the forbidden pairs, then there is a packing of* G *and* H;

(ii) *if* $e(G) + e(H) \le 2n - 4$, *then there is a packing of* G *and* H; and

(iii) *if* $e(G) + e(H) \le [\frac{3}{2}(n - 1)]$, *then there is a packing of* G *and* H.

Proof. It is sufficient to prove (i). Suppose $\Delta(G) \le \Delta(H)$. If $\Delta(H) < n - 1$, then the result follows from Theorem 4.2. If $\Delta(H) = n - 1$ and $\delta(G) = 0$, then we can map an isolated vertex x of G to a vertex y of H such that $d(y) = n - 1$. Then $e(G - x) + e(H - y) \le n - 2$ and one can easily find a packing of $G - x$ with $H - y$. //

Exercise 4.4

1.[+] Suppose G and H are graphs of order n and $e(G) + e(H) \le \frac{3}{2}(n - 2)$. Let $x \in V(G)$, $y \in V(H)$ be such that $d_G(x), d_H(y) \le \frac{1}{2}(n - 2)$. Prove that there is a packing σ of G and H such that $\sigma(x) = y$ (Bollobás and Eldridge [78a]).

(Note that this result has been applied in the study of computational complexity of graph properties. The condition that $d_G(x), d_H(y) \le \frac{1}{2}(n - 2)$ cannot be replaced by $d_G(x) + d_H(y) \le n - 2$. The following is such an example : Let $r \ge 2$, $n \ge 2(r^2 - 1)$, $G = K_{r+1} \cup O_{n-r-1})$, $H = S_{n-r-1} \cup K_{r+1}$, $x \in V(G)$ is a vertex in K_{r+1} and $y \in V(H)$ is the centre of the star S_{n-r-1}. See Bollobás [78;p.425].)

2. Let $r \ge 4$ and let G and H be graphs of order n satisfying

$$\Delta(G) \le \Delta(H) = n - r \quad \text{and} \quad e(G) + e(H) \le 2(n - r) + (r - 1)\sqrt{r}.$$

Prove that if $n \ge 9r^{3/2}$ and there is no packing of G and H, then r is a perfect square, say $r = (k + 1)^2$, and $G = S_{m+1} \cup kK_{k+2} = H$, where $m = n - (k + 1)^2$.

(This is due to S. E. Eldridge, Ph. D. Thesis, University of Cambridge, 1976; see Bollobás [78;p.436].)

3*. Let $k > 2$ be an integer. Prove that any sequence of graphs G_1, G_2, ..., G_k of order n such that $e(G_i) < n - k$ for each $i = 1,...,k$ are packable (Bollobás and Eldridge [78a]).

(Note that Theorem 4.2 implies that this conjecture is true for $k = 2$.)

5. Packing a tree of order n with an (n, n − 1) graph

In the later part of this chapter, we shall generalize the main packing theorem of Bollobás and Eldridge (Theorem 4.2) to a packing of two graphs H and G of order n such that $e(H) + e(G) < 2n - 2$. This generalization is carried out in a few steps. A preliminary step is to show that any tree T of order $n > 5$ can be packed into a graph G of order n and size $n - 1$. This result is due to Slater, Teo and Yap [85]. The original proof of this result uses induction on n by deleting two vertices from both T and G, and as a result we need to verify quite a few cases when $n = 5, 6$. The alternate proof given here is due to Teo [85]. This alternate proof uses induction on n by deleting one vertex from both T and G, and uses the following three lemmas. Lemma 5.1 is due to Hedetniemi, Hedetniemi and Slater [81].

The following definitions are required. A graph of order n and size m is called an <u>(n,m) graph</u>. The tree S'_n of order $n > 5$ is obtained from the star S_{n-1} by inserting a vertex on an edge. The tree $S''_n (n > 6)$ is obtained from S'_{n-1} by inserting a new vertex on the edge which is not incident with the centre of S'_{n-1}. From a given tree T of order $n > 6$, we obtain the trees T(1), T(2), T(3) and a forest T(4) as described below: $T(1) \neq S_{n-1}$ is obtained from T by deleting one end-vertex whose neighbour is of valency at most $n/2$; $T(2) \neq S_{n-2}$ is obtained from T by deleting two end-vertices whose neighbours are distinct; $T(3) \neq S_{n-3}$ is obtained from T by deleting three end-vertices which are not adjacent to one common neighbour; T(4) is obtained from T by deleting a set of three independent vertices which cover at least five edges of T.

Lemma 5.1 *If* T *and* G *are trees of order* $n > 4$, *neither of which is astar, then* T *and* G *are packable.*

Proof. We prove this lemma by induction on n. It is not difficult to verify that this lemma is true for $n < 5$. Hence assume that $n > 6$. We first consider the case $T = S'_n$. Let v be an end-vertex of G, w be the neighbour of v, and x be a vertex of G which is not adjacent to w. Then S'_n can be packed into G by mapping its centre c to v, and a to w where a is the vertex of S'_n which is not adjacent to c. We may now assume that neither T nor G is S'_n, and so each has two end-vertices at distance at least three whose removal does not leave a star. Let t_1 and t_2 be such end-vertices of T with neighbours v_1 and v_2, and choose u_1, u_2, w_1 and w_2 similarly from G. By induction, there is a packing ψ of $T - t_1 - t_2$ into $G - u_1 - u_2$. If $\psi(v_i) \neq w_i$, $i = 1, 2$, then obviously ψ can be extended to a packing of T and G. If $\psi(v_1) = w_1$ and/or $\psi(v_2) = w_2$, then we can rename v_1 by v_2 and vice versa, and ψ is again extendable to a packing of T and G. //

Lemma 5.2 *Let* G *and* H *be two graphs of order* n. *Suppose* G *has an end-vertex* u *whose neighbour is* v *and* H *has an end-vertex* x *whose neighbour is* y. *If* $d_G(v) + d_H(y) < n$ *and there is a packing* π *of* $G - u$ *and* $H - x$, *then there is a packing of* G *and* H.

Proof. If $\pi(v) \neq y$, then π can be extended to a packing of G and H by mapping u to x. Hence we assume that $\pi(v) = y$. Let $A = V(G) - N_G(v)$ and $B = V(H) - N_H(y)$. Since each $a(\neq u) \in N_G(v)$ is such that $\pi(a) \in B$, each $b(\neq x) \in N_H(y)$ is such that $\pi^{-1}(b) \in A$ and $d_G(v) + d_H(y) < n$, there exists $c \in A$ such that $\pi(c) \in B$. Hence the map $\psi : V(G) \to V(H)$ given by $\psi(u) = \pi(c)$, $\psi(c) = x$ and $\psi(d) = \pi(d)$ for every $d \in V(G)$, $d \neq u,c$, is a packing of G and H. //

Although in the proof of Theorem 5.4 we require only part (i) and part (ii) of Lemma 5.3, we include a proof of part (iii) in Lemma 5.3 here because all the three parts will be applied later.

Lemma 5.3 Let T be a tree of order $n \geq 6$. Then

(i) $T(1)$ and $T(2)$ exist if and only if $T \neq S_n$ or S'_n;

(ii) for $n \geq 8$, $T(3)$ exists if and only if $T \neq S_n$, S'_n or P_n; and

(iii) $T(4)$ exists if and only if $T \neq S_n$, S'_n or S''_n.

Proof. For each of the three parts, the necessity is easy to verify, therefore we only need to prove the sufficiency.

(i) Suppose $T(1)$ does not exist. Let x be an end-vertex of T and let y be the neighbour of x. Then either $T - x = S_{n-1}$ or $d(y) \geq \frac{n+1}{2}$. It is clear that $T - x = S_{n-1}$ implies that $T = S_n$ or S'_n. Also if $d(y) \geq \frac{n+1}{2}$ and $T \neq S_n$ or S'_n, then T has at least one branch from y having at least three vertices or T has two branches from y each having at least two vertices. In either case T has another end vertex x_1 such that $T - x_1 = T(1)$.

Suppose $T(2)$ does not exist. Then for any two end-vertices u and v of T having distinct neighbours, $T - u - v = S_{n-2}$. Since $n \geq 6$, either u or v must be adjacent to the centre of S_{n-2} and so $T = S'_n$.

(ii) Suppose $T \neq P_n$. Then T has at least three end-vertices x_1, x_2 and x_3. Let their neighbours be y_1, y_2 and y_3 respectively. In each of the three cases $|\{y_1, y_2, y_3\}| = 1, 2, 3$, it is not difficult to show that if $T \neq S_n$ or S'_n, then $T(3)$ exists.

(iii) If $T(4)$ does not exist, then the diameter of T is at most 4 and it is easy to see that $T = S_n$, S'_n or S''_n. //

Theorem 5.4 Let T be a tree of order $n \geq 5$ and let G be an $(n, n - 1)$ graph. If neither T nor G is a star, then there is a packing of T and G.

Proof. By Lemma 5.1, we can assume that G is not a tree. We now prove this theorem by induction on n.

Claim 1. The theorem is true for $n = 5$.

Since G is not a tree, G contains a cycle. Also, since G is an $(n, n - 1)$ graph which is not a tree, G is not connected. Hence

$G = K_2 \cup K_3$, $O_1 \cup C_4$ or $O_1 \cup G_1$ where G_1 is obtained from K_4 by deleting two adjacent edges. In any of these three cases, it is not difficult to verify that $T = P_5$ or S_5' (note that if $T \neq S_5$, then $T = P_5$ or S_5') can be packed into G.

Claim 2. The theorem is true for $T = S_n'$.

Since G is an (n, n - 1) graph where $n \geq 5$, either G has an isolated vertex or G has an end-vertex. If G has an isolated vertex, it is clear that T and G are packable. Suppose G has an end-vertex, v_1 say. Then $G - v_1 \neq K_{n-1}$. Let v_2 be the neighbour of v_1 and $v_2v_3 \notin G$. Then we can map the centre c of S_n' to v_1 and the vertex of S_n' which is not adjacent with c to v_2 to obtain a packing of S_n' and G.

Claim 3. The theorem is true if G has an isolated vertex.

Suppose v_1 is an isolated vertex of G. By deleting a suitable edge e from $G - v_1$, we obtain a graph $G' = G - v_1 - e$ such that $\Delta(G') < n - 2$ and G' is an (n - 1, n - 2) graph. Suppose T is a path. Let π be a packing of T(1) into G'. If T(1) overlaps with $e = v_2v_3$ (say) of G', then we can modify π to a packing of T into G by replacing e by v_2v_1 and v_1v_3. If T(1) does not overlap with e, then we can extend π to a packing of T into G by joining v_1 to an end-vertex of T(1). Hence we assume that T is not a path. Then T has a vertex x of valency at least three. We now have $e(T - x) + e(G - v_1) \leq (n - 4) + (n - 1) = 2(n - 1) - 3$. By Theorem 4.2, if $\Delta(G - v_1) < n - 2$, then there is a packing π of T - x into $G - v_1$ unless $T - x = 3K_2$ and $G - v_1 = O_2 \cup K_4$. It is clear that π can be extended to a packing of T into G. If $T - x = 3K_2$ and $G - v_1 = O_2 \cup K_4$, then T and G are shown in Fig.4.4 and in this case it is easy to see that T can be packed into G. On the other hand, if $\Delta(G - v_1) = n - 2$, then G is the graph shown in Fig.4.5 and in this case it is also easy to see that T can be packed into G.

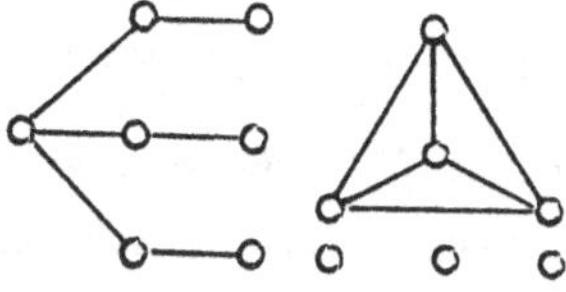

Figure 4.4

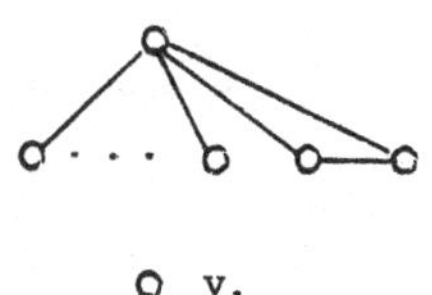

Figure 4.5

Claim 4. The theorem is true if G has no isolated vertices.

If G has no isolated vertices, then G has at least one end-vertex. By Claim 2, we assume that $T \neq S'_n$. By Lemma 5.2 and induction, we may assume that $G = S_k \cup C_{r_1} \cup \ldots \cup C_{r_t}$, where $k > \frac{n}{2} + 1$ and $r_1 > \ldots > r_t > 3$. If G has a cycle-component C_m, $m > 4$, we can replace it by C_{m-1} and obtain an $(n - 1, n - 2)$ graph G'. By induction, there is a packing π of $T(1) = T - x$ with G'. Let y be the neighbour of x. We can now modify π to a packing of T into G by mapping x to a vertex of C_m so that xy does not overlap with an edge of C_m. Hence we assume that all the components of G are triangles. Also, by induction, we can assume that $G = S_{n-3} \cup C_3$ (otherwise since $n > 8$, we may pack T', where T' is obtained from T by deleting three independent vertices, into $G - C_3$ and extend it to a packing of T into G). Now since $k > \frac{n}{2} + 1$, we have $n > 8$. Thus by Lemma 5.3 (ii), if $T \neq P_n$, then T has three end-vertices x, y and z such that the neighbour w of x does not belong to $N(y) \cup N(z)$. Let $T(3) = T - \{x, y, z\}$. Then we can obtain a packing of T into G by mapping y, z and w to C_3 and x on the centre of S_{n-3}. Finally if $T = P_n$, it is easy to see that T can be packed into $G = S_{n-3} \cup C_3$. //

Exercise 4.5

1. Let P'_6 be the tree shown below

P'_6 :

Prove that there does not exist a packing of three copies of P'_6 into K_6 (Huang and Rosa [78]).

2. For $n > 7$, let P'_n be the tree shown below

P'_n :

Does there exist a packing of three copies of P'_n into K_n?

3. Suppose r and s are integers such that $r < s < n$. Prove that any three trees T_r, T_s and T_n can be packed into K_n (due to A.M. Hobbs and B. Bourgeois, see Hobbs [81]).

4. Let T_n, T'_n and T''_n be any trees of order $n \geq 7$ such that $\Delta(T_n)$, $\Delta(T'_n)$, $\Delta(T''_n) \leq n - 3$. Does there exist a packing of T_n, T'_n and T''_n into K_n?

6. Packing a tree of order n with an (n,n) graph

In this section we shall find all the pairs $\{T,G\}$ where T is a tree of order $n \geq 5$ and G is an (n,n) graph such that T cannot be packed into G. This result (Theorem 6.2) extends Theorem 5.4. However, in the proof of Theorem 6.2 we need to apply Theorem 5.4 and some previous results. We also need the following lemma which settles some special cases of Theorem 6.2.

Lemma 6.1 Suppose T is a tree of order $n \geq 5$ and G is an (n,n) graph such that $\Delta(T), \Delta(G) < n - 1$.

(i) If $T = S'_n$ and $G \neq \cup C_i$, then there is a packing of T and G.

(ii) If $n \geq 7$ and G has two vertices u_1 and u_2 such that $e(G - u_1 - u_2) \leq 1$, then there is a packing of T and G.

(iii) If $G = \cup C_i$ where $i \geq 4$ for at least one i and $T \neq S'_n$ or if $G = kC_3$ and $T \neq S'_n$ or S''_n, then there is a packing of T and G.

(iv) If G is obtained from $G' = S_k \cup (\cup C_i)$, $k \geq 4$, by adding an edge joining two non-adjacent vertices of G' none of which is the centre of S_k, then there is a packing of T and G.

Proof. (i) By assumption, G has an isolated vertex or an end-vertex. The proof of this result is similar to that of Claim 2 in Theorem 5.4.

(ii) By (i) we can assume that $T \neq S'_n$. Hence, by Lemma 5.3(i), T has two end-vertices x_1 and x_2 such that $T(2) = T - x_1 - x_2$ is not a star. Let $N(x_1) = \{x_3\}$, $N(x_2) = \{x_4\}$, and let $G' = G - u_1 - u_2$. We first observe that G has two vertices u_3 and u_4 such that u_1u_3, u_2u_4, $u_3u_4 \notin G$, for otherwise either G has at least $n - 3$ vertices each of which is adjacent to both u_1 and u_2, and so $r(n - 3) \leq e(G) \leq n$, from which it follows that $n \leq 6$, a contradiction; or G has $n - 2$ vertices each of which is adjacent to both u_1 and u_2, and the remaining two vertices form a K_2-component of G, from which it follows that $G = K_2 \cup K_{3,2}$ and in this case T and G are packable; or $d(u_1) = n - 2$ and $d(u_2) < 2$ and thus G has another vertex $u_2' \neq u_2$ so that u_1 and u_2' play the role of u_1 and u_2. We now define a map $\sigma : V(T) \to V(G)$ by setting $\sigma(x_i) = u_i$ for $i = 1, 2, 3, 4$. If $G' = O_{n-2}$, it is clear that σ can be extended to a packing of T and G. If $G' = O_{n-4} \cup K_2$, since $T' = T - \{x_1, x_2, x_3, x_4\}$ is a forest of order at least 3, we can map two non-adjacent vertices of T' on K_2 to extend σ to a packing of T and G.

(iii) It is not difficult to verify this result for $n = 5, 6$. (A list of all trees of order 5 and 6 can be found in Harary [69; p.233].) Hence we assume that $n \geq 7$. Suppose G contains C_m for some $m \geq 4$. Let $C_m = v_1v_2 \dots v_mv_1$. Since $T \neq S_n$ or S_n', T(2) exists (by Lemma 5.3(i)), and by Theorem 5.4, there is a packing of T(2) and $G - v_1 - v_2$, which can be extended to a packing of T and G. Hence we assume that $G = kC_3$, $k \geq 3$. Since $T \neq S_n$, S_n' or S_n'', T(4) exists (by Lemma 5.3(iii)). Hence, by Theorem 4.2, there is a packing of T(4) and $G - C_3$, which can be extended to a packing of T and G.

(iv) Suppose e is an edge joining two vertices belonging to two distinct components of G' or two non-adjacent vertices of C_i for some $i \geq 5$, or two vertices of S_k. By Theorem 5.4, there is a packing of T and G'. By the symmetry of G' and the fact that T contains no cycles, we can pack T with G' so that e does not overlap with any edge of T. Hence T and G are packable.

Finally, suppose e is an edge joining two opposite vertices v_2 and v_4 of $C_4 = v_1v_2v_3v_4v_1$. Let $G'' = G - v_1$ and let x be an end-vertex of T such that $T - x \neq S_{n-1}$. By Theorem 5.4, there is a packing σ of $T - x$ and G". Suppose the neighbour of x is y. By the symmetry of a triangle, we can assume that $\sigma(y) \neq v_2, v_4$ and so σ can be extended to a packing of T and G by mapping x to v_1. //

We shall need the following notation. Let C_4^+ be the graph obtained from C_4 by adding an edge joining two opposite vertices and let S(6) be the spider obtained from S_4 by adding two new vertices each of which is joined to one end-vertex of S_4.

Theorem 6.2 *Suppose* T *is a tree of order* $n > 5$ *and* G *is an* (n,n) *graph such that* $\Delta(T), \Delta(G) < n - 1$. *If* $\{T,G\} \neq \{P_5, O_1 \cup C_4^+\}, \{P_6, O_2 \cup K_4\}, \{S(6), O_2 \cup K_4\}$, *or if* $G = \cup C_i$ *where* $i > 4$ *for at least one* i *and* $T \neq S'_n$ *or if* $G = kC_3$ *and* $T \neq S'_n$ *or* S''_n, *then there is a packing of* T *and* G.

Proof. We first prove that this theorem is true if G is connected. It is not difficult to verify this for $n = 5$. (A list of connected (5,5) graphs can be found in Harary [69; p.216].) Hence we assume that $n > 6$. If $G = C_n$, then by Lemma 6.1(iii), T and G are packable unless $T = S'_n$. Suppose $G \neq C_n$. Now by Lemma 6.1(i), we assume that $T \neq S'_n$. Since G is connected and $G \neq C_n$, G has an end-vertex u. Let v be the neighbour of u. We consider two cases.

Case 1. $d(v) < n/2$.

Since $T \neq S_n$ or S'_n, by Lemma 5.3(i), T(1) exists. If $G - u \neq C_{n-1}$ and $\Delta(G - u) < n - 2$, then by induction, there is a packing of T(1) and $G - u$. Thus, by Lemma 5.2, T and G are packable.

Suppose $G - u = C_{n-1}$. Then by Theorem 5.4, there is a packing σ of T and $G - uv$. Since $T \neq S_n$ or S'_n, by the symmetry of C_{n-1}, we can assume that no edge of T lies on the edge uv in the packing σ, and so σ is also a packing of T and G.

Suppose $\Delta(G - u) > n - 2$. If $n = 6$, then G is the graph given in Fig.4.6 and we can verify that T and G are packable. If $n > 7$, then by Lemma 6.1(ii), T and G are packable.

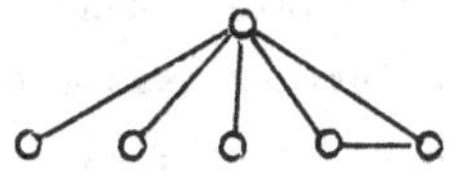

Figure 4.6

Case 2. $d(v) > n/2$.

If G has two end-vertices u_1 and u_2 with distinct neighbours v_1 and v_2 such that $d(v_1), d(v_2) > n/2$, then the case that $n > 7$ is settled by Lemma 6.1(ii), and the case that $n = 6$ can be verified directly. Hence we assume that any end-vertex of G is adjacent to v. Since G is connected, G has exactly one cycle. Hence $G = G_1$ or G_2, where G_1 and G_2

are as shown in Fig.4.7.

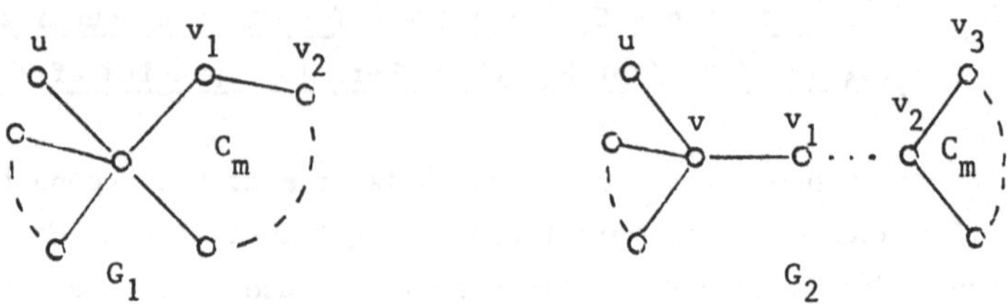

Figure 4.7

Suppose $G = G_1$. If $n = 6$, then G is as given in Fig. 4.8(a) and we can verify that T and G are packable. Hence we assume that $n \geq 7$ and by Lemma 6.1(ii), we can further assume that $m \geq 6$. Since $T \neq S_n$ or S'_n, by Lemma 5.3(i), T(2) exists and thus by Theorem 5.4, there is a packing σ of T_2 and $G_1 - v_1 - v_2 \neq S_{n-2}$. This packing can be extended to a packing of T and G_1.

Figure 4.8

Suppose $G = G_2$. If $n = 6$, then G is as given in Fig. 4.8(b) and we can verify that T and G are packable. Hence we assume that $n \geq 7$ and by the previous argument we can further assume that $m = 3$ and that the length of the path from v to v_2 is at least 3. Hence $n \geq 8$. If $T \neq P_n$, then by Lemma 5.3(ii), T(3) exists, and by Theorem 5.4, there is a packing σ of T_3 and $G' = G_2 - v_2 - v_3 - v_4 \neq S_{n-3}$, where v_3 and v_4 are the other two vertices of the triangle. This packing σ can be extended to a packing of T and G_2. If $T = P_n$, let u_1 and u_2 be the two end-vertices of T and let u_3 be a middle vertex of T which is not adjacent to u_1 or u_2. Then any packing of $T - u_1 - u_2 - u_3$ and G' can be extended to a packing of T and G_2.

We next prove that this theorem is true if G is not connected.

Case (i). G has an isolated vertex u.

Let $v \in G$ be such that $d(v) = \Delta(G) \geq 3$. Then $G' = G - u - v$ is an $(n - 2, q)$ graph, where $q \leq n - 3$ and $\Delta(G') < n - 3$ if $n \geq 7$. Let x be an end-vertex of T such that its neighbour y has maximum valency among all the vertices which are adjacent to the end-vertices of T. Then $T - x - y \neq S_{n-2}$. Hence if $n \geq 7$, then by Theorem 5.4, there is a packing σ of $T - x - y$ and G'. This packing can be extended to a packing of T and G by mapping x to v and y to u. If $n = 5$, then G has only one isolated vertex. Thus $G = O_1 \cup C_4^+$ and if $T \neq P_5$, T and G are packable. If $n = 6$ and G has only one isolated vertex, then it is not difficult to verify that T and G are packable for any T. (A list of (6,6) graphs having exactly one isolated vertex can be found in Harary [69; p.217].) If $n = 6$ and G has two isolated vertices, then $G = O_2 \cup K_4$ and so T and G are packable except when $T = P_6$ or $S(6)$.

Case (ii). G has no isolated vertices.

We can verify that this theorem is true for $n = 5$ and 6. (There are no disconnected (5,5) graphs having no isolated vertices and there are only two disconnected (6,6) graphs having no isolated vertices.) Hence we assume that $n \geq 7$.

Suppose G has a tree-component. It is clear that if $T = S_n'$, then T and G are packable. Hence we assume that $T \neq S_n'$ and so by Lemma 5.3(i), T(1) exists. Let u be an end-vertex of G and let v be the neighbour of u. Then, by induction, T(1) and $G - u$ are packable. Now if $d(v) < n/2$, then by Lemma 5.2, T and G are packable. Hence we assume that $d(v) \geq \frac{n}{2}$ and thus the tree-component of G is S_k, where $2k \geq n$ and thus $k \geq 4$. Hence $G - S_k$ is the union of cycles with an additional edge joining two vertices of $G - S_k$. However, by Lemma 6.1(iv), T and G are also packable in this special case.

Suppose that all the components of G are (n_i, n_i) graphs. By Lemma 6.1(iii), we can assume that G has at least one component which is not a cycle. Such a component has an end-vertex. If G has at least two components which are not cycles, then one of these components has an end-vertex u whose neighbour v is such that $d(v) < n/2$. In this case, it should be clear, by now, that T and G are packable. Hence we assume that G has only one component H which is not a cycle and all the other components of G are cycles. Also, since G has an end-vertex, we may further assume that $T \neq S'_n$. Suppose G has a cycle $v_1 v_2 \ldots v_m v_1$ where $m > 4$. Then any packing of T(2) with $G - v_1 - v_2$ can be extended to a packing of T and G. Hence we assume that all the cycle-components of G are triangles, i.e. $G = H \cup kC_3$. In this case, it is not difficult to see that if $T = S''_n$, then T and G are packable. On the other hand, if $T \neq S''_n$, then by Lemma 5.3(iii) and Theorem 4.2, there is a packing of T(4) and $G - C_3$ which can be extended to a packing of T and G. //

Corollary 6.3 *Suppose* T *is a tree of order* $n > 7$ *and* G *is an* (n,n) *graph* such that $\Delta(T), \Delta(G) < n - 1$. *Then* T *and* G *are packable except* (i) $G = \cup C_i$ *where* $i > 4$ *for at least one* i *and* $T = S'_n$; *or* (ii) $G = kC_3$ *and* $T = S'_n$ *or* S''_n.

Exercise 4.6.

1. Find a tree T of order $n > 5$ and an (n, n + 1) graph G such that (i) $\Delta(T), \Delta(G) < n - 1$; (ii) $\{T,G\}$ is neither one of the forbidden pairs given in Theorem 6.2 nor $\{T,G'\}$ is one of the forbidden pairs given in Theorem 6.2 and G is obtained from G' by adding an edge; and (iii) T cannot be packed with G.

2. Does there exist an infinite family of pairs $\{T,G\}$ where T is a tree of order $n > 7$, G is an (n, n + 1) graph such that (i) $\Delta(T), \Delta(G) < n - 1$; (ii) $\{T,G\}$ is neither one of the forbidden pairs given in Corollary 6.3 nor $\{T,G'\}$ is one of the forbidden pairs given in Corollary 6.3 and G is obtained from G' by adding an edge; and (iii) T cannot be packed with G?

7. Packing two (n, n - 1) graphs

Burns and Schuster [78] proved that if G is an (n, n - 1) graph where $n > 6$ if n is even and $n > 9$ if n is odd, then G is embeddable in $\overline{G}$ if and only if G is neither S_n nor $K_3 \cup S_{n-3}$ $(n > 8)$. Sauer and Spencer [78] proved that any two (n, n - 2) graphs are packable. In this section we shall prove that if G and H are two (n, n - 1) graphs, $n \geq 5$, which are not stars, then except for 12 forbidden pairs for $n \leq 11$ and for an infinite family where $G = K_3 \cup S_{n-3} = H$, G and H are packable. This theorem (due to Teo and Yap [-a]) generalizes the above mentioned results and Theorem 5.4. From this theorem, it follows that if $n \geq 12$, then any two (n, n - 1) graphs G and H which are not stars, are packable unless $G = K_3 \cup S_{n-3} = H$.

To prove this, we need to apply Theorem 5.4 and Lemma 7.1. We also need the following notation. Suppose G is a graph and $\overline{G}$ contains C_r. Then, by abusing the notation, the subgraph of G induced by $V(G) - V(C_r)$ is denoted by $G - C_r$.

Lemma 7.1 <u>Let</u> G <u>be an</u> (n, n - 1) <u>graph</u>, $n \geq 6$, <u>which is not a star. Suppose</u> $G \neq K_3 \cup S_{n-3}$ $(n \geq 7)$, $O_3 \cup K_4$ <u>or</u> $O_1 \cup 2K_3$. <u>Then</u> $\overline{G}$ <u>contains</u> C_r <u>for each</u> $r = 3, 4, \ldots, n - 2$ <u>such that</u> C_r <u>covers at least</u> $r + 1$ <u>edges of</u> G <u>and</u> $G - C_r$ <u>has an isolated vertex</u>.

Proof. Let A be the set of isolated vertices of G, let B be the set of nontrivial tree-components of G, and let C be the set of other components of G.

We first prove that this lemma is true for r = 3. If $C = \phi$, then G is a tree. Suppose G is a path given by $x_1x_2 \ldots x_n$, then $\{x_2, x_4, x_6\}$ is a required C_3. If G is not a path, then since $G \neq S_n$, G has three end-vertices u, v and w such that u and v are not adjacent to x where x is the neighbour of w. It is clear that $\{u, v, x\}$ is a required C_3. Hence we assume that $C \neq \phi$. We consider two cases separately.

Case 1. $A = \phi$.

We need only to consider the case $|B| = 1 = |C|$, because if $|B| + |C| \geq 3$, then we can choose three vertices for C_3 from three components of G. Let $B = \{T\}$ and let $C = \{H\}$ where H is a (t,t) graph. If $t \geq 4$ or if $H = K_3$ and $T \neq S_{n-3}$, $n \geq 7$, then it is easy to find a required C_3.

Case 2. $A \neq \phi$.

We first note that $A \neq \phi$ implies that $C \neq \phi$. If $B \neq \phi$, then

$\Delta(H) \geq 3$ for at least one $H \in C$ and we can choose $x \in A$, $y \in V(T)$ where $T \in B$, and $z \in V(H)$ such that $d_H(z) \geq 3$, for the three vertices of a required C_3. Hence, from now on, we assume that $B = \phi$.

Suppose $|A| \geq 2$. Then the lemma is obviously true if either $|C| \geq 2$ or if $C = \{H\}$ and H is not a complete graph. On the other hand, if $H = K_s$, then since G is an $(n, n - 1)$ graph, $s \geq 4$. If $s = 4$, then $G = O_3 \cup K_4$ is a forbidden graph. If $s \geq 5$, then $|A| \geq 3$ and we can choose two vertices from A and one vertex from K_s to form a required C_3.

Suppose $|A| = 1$. If $|C| \geq 2$, then the lemma is true unless $|C| = 2$ and the two graphs in C are triangles, i.e. $G = O_1 \cup 2K_3$, a forbidden graph again.

Finally we consider the case $|A| = 1 = |C|$. Let $C = \{H\}$. For $n = 6$, we can verify that $O_1 \cup H$ has a required C_3. (There are five connected (5,5) graphs, see Harary [69; p.216].) Hence we assume that $n \geq 7$. Now, by Theorem 4.2, C_3 can be embedded in $\overline{H}$. If we cannot find an embedding of C_3 in $\overline{H}$ so that C_3 covers at least four edges of H, then C_3 covers exactly three edges of H in such a way that these three edges are all incident with the same vertex, x say. Now each of the other vertices in $H - (C_3 \cup \{x\})$ must be of valency one and all of them are incident with x also. Hence $H = S_{n-1}$, contradicting the fact that H is an $(n - 1, n - 1)$ graph.

We next prove that this lemma is true for $4 \leq r \leq n - 2$.

Let $V(G) = \{x_1, x_2, \ldots, x_n\}$ with

$$d(x_1) \geq d(x_2) \geq \ldots \geq d(x_k) \geq d(x_{k+1}) \geq \ldots \geq d(x_n),$$

where $d(x_k) \neq 0$ and $d(x_{k+1}) = 0$. Let $D = \{x_1, \ldots, x_k\}$ and let H be the subgraph of G induced by D.

Case (i). $d(x_k) \geq 4$.

In this case, $|A| \leq (n+1)/2$. For any $3 \leq r \leq n - 2$, we have $[\frac{r}{2}] + 1 \leq n/2$. Thus we can choose $[\frac{r}{2}]$ vertices from D (if $|D| \geq [\frac{r}{2}]$) and $[\frac{r+1}{2}]$ vertices from A; or $|D|$ vertices from D (if $[\frac{r}{2}] > |D|$) and $r - |D|$ vertices from A, to form a required C_r.

Case (ii). $d(x_k) = 3$.

In this case, $|A| \geq \frac{n+2}{3} \geq 3$. Let $u \in A$, let $w = x_k$ and let $e = vw \in G$. Then $G' = G - u - e$ is an $(n - 1, n - 2)$ graph which is not a star. Since G' contains at least two isolated vertices and $d_{G'}(w) = 2$, G' cannot be a forbidden graph. By induction (we can check through the list of (6,5) graphs given in Harary [69; p.219] to verify that the lemma is true for all (6,5) graphs), G' contains C_r, $r = 3, \ldots, n - 3$, so that C_r covers at least $r + 1$ edges of G' and $G' - C_r$ has an isolated vertex. If $w \in C_r$, then C_r can be turned into a required C_{r+1} for G by joining u to w and z where $wz \in C_r$ and deleting wz from C_r. Similarly, if $v \in C_r$, we can also obtain a required C_{r+1} for G. Now if $w, v \notin C_r$ but $r \geq 5$, then since $d_{G'}(w) = 2$, w is not adjacent to at least two neighbouring vertices, x and y say, in C_r and so we can turn C_r into a required C_{r+1} for G by joining w to x and y and deleting xy from C_r. It remains to show that $\overline{G}$ contains a required C_4 and C_5. Let $a, b \in A$. Then wavbw forms a required C_4. Also if $|A| \geq 4$, then for $a, b, c \in A$, $x, y \in D$, xaybcx forms a required C_5. The case that $|A| = 3$ leads to $n = 7$ and $G = O_3 \cup K_4$, a forbidden graph.

Case (iii). $d(x_k) \leq 2$.

Suppose $A \neq \phi$. Let $w = x_k$, $e = wv \in G$ and let $G' = G - a - e$, $a \in A$. Then G' is an $(n - 1, n - 2)$ graph which is not a star and is not a forbidden graph. (This is always possible because $d(w) \leq 2$.) By induction, $\overline{G'}$ contains C_r, for $r = 3,4,\ldots,n - 3$ satisfying the conditions of the lemma. Then as in the proof of Case (ii), we can turn C_r into a required C_{r+1} in $\overline{G}$.

Suppose $A = \phi$. Then G contains at least one end-vertex x and $G' = G - x$ is an $(n - 1, n - 2)$ graph. We note that since $G \neq S_{n-3} \cup K_3$, G has an end-vertex x so that G' is not a forbidden graph. If $n \geq 7$, then by induction, $\overline{G'}$ contains C_r, $r = 3,4,\ldots,n - 3$ satisfying the conditions of the lemma. Each C_r can be turned into a required C_{r+1} in $\overline{G}$ by joining x to two neighbouring vertices of C_r where none of these two neighbouring vertices is the neighbour of x.

Finally, we can verify that the lemma is true for $r = 4$ and $n = 6$. (A list of (6,5) graphs G such that G contains no isolated vertices can be found in Harary [69; p.219].) //

The graphs G(5), H_7 and G(8) are depicted in Fig.4.9.

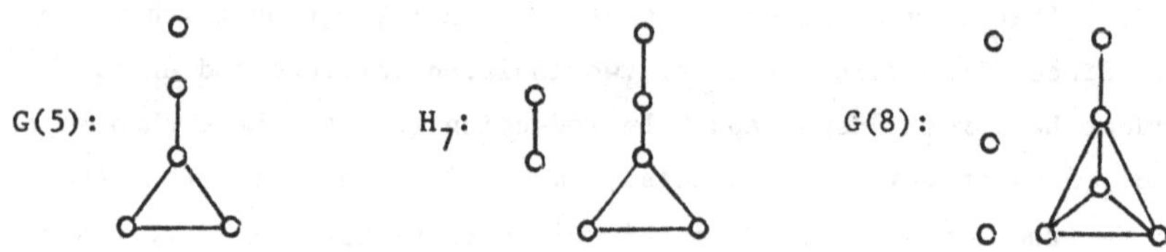

Figure 4.9

Theorem 7.2 Suppose G and H are two (n, n - 1) graphs, $n \geq 5$, which are not stars. If $\{G,H\}$ is not one of the following thirteen pairs:

(1) $\{P_2 \cup K_3, P_2 \cup K_3\}$, (2) $\{O_1 \cup C_4, O_1 \cup C_4\}$, (3) $\{G(5), P_2 \cup C_3\}$,
(4) $\{O_3 \cup K_4, P_2 \cup C_5\}$, (5) $\{O_3 \cup K_4, P_4 \cup K_3\}$, (6) $\{O_3 \cup K_4, H_7\}$,
(7) $\{O_1 \cup 2K_3, S_4 \cup K_3\}$, (8) $\{O_1 \cup 2K_3, O_3 \cup K_4\}$,
(9) $\{O_1 \cup 2K_3, O_1 \cup 2K_3\}$, (10) $\{G(8), P_2 \cup 2K_3\}$,
(11) $\{O_2 \cup P_2 \cup K_4, P_2 \cup 2K_3\}$, (12) $\{O_6 \cup K_5, P_2 \cup 3K_3\}$, and
(13) $\{K_3 \cup S_{n-3}, K_3 \cup S_{n-3}\}$, $n \geq 8$,

then there is a packing of G and H.

Proof. By Theorem 5.4, we can assume that both G and H are not connected. For n = 5, we can verify that there are three forbidden pairs given by (1), (2) and (3). (There are three (5,4) graphs which are not connected, see Harary [69; p.216].) Hence we assume that $n \geq 6$. Let F be the forbidden pairs given in Theorem 4.2.

Case 1. G has an isolated vertex u.

(i) Suppose $\Delta(H) \geq 3$.

Let $v \in V(H)$ be such that $d(v) = \Delta(H)$. By the choice of v, $\Delta(H - v) < n - 2$. Hence, if $\Delta(G - u) < n - 2$ and $\{G - u, H - v\} \notin F$, then (since $e(G - u) + e(H - v) \leq 2(n - 1) - 3$), by Theorem 4.2, there is a packing of G - u and H - v, which can be extended to a packing of G and H. If $\{G - u, H - v\} \in F$, then $\{G - u, H - v\} = \{O_2 \cup K_4, 3K_2\}$, $\{2K_3, O_3 \cup K_3\}$ or $\{3K_3, O_5 \cup K_4\}$ and we deduce that the forbidden pairs for $\{G, H\}$ are given by (6), (7) and (8). On the other hand, if

$\Delta(G - u) = n - 2$, let $w \in G$ be such that $d(w) = n - 2$. Then G is the graph given in Fig.4.10 below. In this case, if H has an isolated

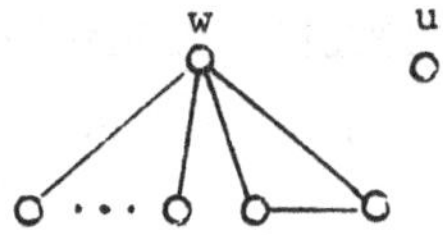

Figure 4.10

vertex x, we can map w to x to obtain a packing of G and H; otherwise H has an end-vertex y and we can map w to y, u to z (z is the neighbour of y), to obtain a packing of G and H.

(ii) Suppose $\Delta(H) = 2$.

In this case, H is the union of some cycles and a path. Since G has an isolated vertex, by Lemma 7.1, if $G \neq O_1 \cup 2K_3$ or $O_3 \cup K_4$, we can assume that $\overline{G}$ contains C_r for each $r = 3, 4, \ldots, n - 2$ such that C_r covers the maximum number of edges of G (which is at least $r + 1$) and $G - C_r$ has an isolated vertex. Suppose $H \neq O_1 \cup C_{n-1}$. Let C_r be the smallest cycle-component of H. Then $e(G - C_r) + e(H - C_r) \leq (n - 1) - (r + 1) + (n - 1) - r = 2(n - r) - 3$. Hence if $\Delta(H - C_r) < n - r - 1$, i.e. $H - C_r \neq P_2$ or P_3, and if $\{G - C_r, H - C_r\} \notin F$, then by Theorem 4.2, $G - C_r$ and $H - C_r$ are packable, from which it follows that G and H are packable. However, if $H - C_r = P_2$ or P_3, then $r = n - 2$ or $n - 3$ and $e(G - C_r) = 0$ or 1. Thus G and H are also packable. Next, if $\{G - C_r, H - C_r\} \in F$, then by Theorem 4.2, it must be either $\{O_2 \cup K_3, K_2 \cup K_3\}$ or $\{O_4 \cup K_4, K_2 \cup 2K_3\}$. Hence, in either case $r = 3$, and $n = 8$ or 11.

Let $V(C_3) = \{a,b,c\}$. For $n = 8$, let the two isolated vertices of $G - C_3$ be v and w, and let the three vertices of the triangle in $G - C_3$ be x, y and z. Our aim now is to add four edges joining $\{a,b,c\}$ to $\{v,w,x,y,z\}$ so that we can single out the forbidden pairs for $\{G, H\}$. By symmetry, we need only to discuss the following three possibilities:

(i) $ax, ay, az, av \in G$ yield the forbidden pair (10).

(ii) $ax, ay, az, bv \in G$ yield the forbidden pair (11).

(iii) In each of the other cases we can embed C_3 in $\overline{G}$ so that C_3 covers more than four edges of G, a contradiction to the assumption that C_r covers the maximum number of edges of G. (We call this kind of argument <u>a maximal covering argument</u>.)

For n = 11, applying the maximum covering argument, we find only one forbidden pair (12).

Next, suppose $G = O_3 \cup K_4$. Then n = 7 and $H = O_1 \cup 2K_3$, $P_2 \cup C_5$, $P_3 \cup C_4$ or $P_4 \cup C_3$. In this case we can easily obtain the forbidden pairs (4), (5) and (8). If $G = O_1 \cup 2K_3$, then we obtain the forbidden pair (9). Also if $H = O_1 \cup C_{n-1}$, then by interchanging the role of G and H, we can assume that $G = O_1 \cup C_{n-1}$ and thus G and H are packable.

Case 2. Both G and H have no isolated vertices.

Since both G and H are not connected, each of them has at least one tree-component. Suppose G has no end-vertex u whose neighbour u' is such that $d(u') < n/2$. Then $G = S_t \cup C_{r_1} \cup \ldots \cup C_{r_i}$, where $t \geq \frac{n}{2} + 1$, $3 \leq r_1 \leq r_2 \leq \ldots \leq r_i$. Hence $n \geq 8$. By Lemma 7.1, if $H \neq S_{n-3} \cup K_3$, then $\overline{H}$ contains C_{r_1} such that $H - C_{r_1}$ has at most $n - (r_1 + 1) - 1 = n - r_1 - 2$ edges and has an isolated vertex. Hence, by Theorem 4.2, if $G - C_{r_1} \neq S_t$, then since $\{G - C_{r_1}, H - C_{r_1}\} \notin F$, there is a packing of $G - C_{r_1}$ and $H - C_{r_1}$. Also, if $G - C_{r_1} = S_t$, then since $H - C_{r_1}$ has an isolated vertex, there is also a packing of $G - C_{r_1}$ and $H - C_{r_1}$. In either case, the packing of $G - C_{r_1}$ and $H - C_{r_1}$ can be extended to a packing of G and H. On the other hand, if $H = S_{n-3} \cup K_3$ and $G \neq S_{n-3} \cup K_3$, then by interchanging the role of G and H in the above argument, we can see that G and H are packable. This shows that if G has no end-vertex u whose neighbour u' is such that $d(u') < n/2$, then there is always a packing of G and H. The case that $G = S_{n-3} \cup K_3 = H$, $n \geq 8$ yields the forbidden pair (13).

By the previous discussion, we can now assume that G (resp. H) has an end-vertex u (resp. v) whose neighbour u' (resp. v') is such that $d_G(u') < n/2$ (resp. $d_H(v') < n/2$). Thus, by Lemma 5.2, if G - u and H - u are packable, then G and H are packable. It remains to consider the case that $G - u$ and $H - v$ are not packable. We distinguish two cases.

(i) At least one of $G - u$ and $H - v$ has an isolated vertex.

Suppose $G - u$ has an isolated vertex. Since G has no isolated vertex, $G - u$ has exactly one isolated vertex and thus G has K_2 as a component. Now by examining the forbidden pairs obtained in Case 1 (note that H has no isolated vertex and $H - v$ has at most one isolated vertex), $\{G - u, H - v\}$ must be one of the forbidden pairs (2), (3), (7) and (9).

If $\{G - u, H - v\}$ is the forbidden pair (2), then $G = K_2 \cup C_4 = H$ and it is easy to see that G and H are packable. If $\{G - u, H - v\}$ is the forbidden pair (3), then $H = P_3 \cup C_3$, and G and H are packable. If $\{G - u, H - v\}$ is the forbidden pair (7), then $G = K_2 \cup 2K_3$ and $H = S_5 \cup K_3$, and G and H are packable. If $\{G - u, H - v\}$ is the forbidden pair (9), then $G = K_2 \cup 2K_3 = H$, and G and H are packable.

(ii) Both $G - u$ and $H - v$ have no isolated vertices.

By the previous argument and by induction, G and H are always packable unless $G - u = S_{n-4} \cup K_3 = H - v$ and thus $G = S_{n-3} \cup K_3 = H$.

The proof of Theorem 7.2 is complete. //

Exercise 4.7

1. Applying Theorem 7.2, show that if G is an $(n, n - 1)$ graph where $n \geq 6$, then G is embeddable in $\overline{G}$ if and only if G is neither S_n nor $K_3 \cup S_{n-3}$ $(n \geq 8)$.

2. Applying Theorem 7.2, show that any two $(n, n - 2)$ graphs are packable.

3. Characterize the (n, n) graphs G, $n \geq 5$, such that G can be packed into its complement (Faudree, Rousseau, Schelp and Schuster [79]).

4. Suppose G is a graph of order n. Prove that if $e(G) \leq \frac{6}{5}n - 2$, $G \neq S_n$ and G contains no cycles of length 3 or 4, then G can be packed into its complement (Faudree, Rousseau, Schelp and Schuster [79]).

5.* Prove that every non-star graph which contains no cycles of length 3 or 4 can be packed into its complement (Faudree, Rosseau, Schelp and Schuster [79]).

8. Packing two graphs of order n having total size at most 2n − 2

In this section, we shall apply the results of sections 4 to 7 to find all the forbidden pairs $\{G, H\}$ where G and H are graphs of order $n \geq 5$ such that $\Delta(G), \Delta(H) < n - 1$, $e(G) + e(H) \leq 2n - 2$, and G and H are packable. This result (due to Teo and Yap [-b]) extends Theorem 4.2.

We shall need the following lemma whose proof is similar to that of Lemma 4.1.

Lemma 8.1 *Let* T *be a tree of order* p *and let* G *be a graph of order* n. *Suppose* $4 \leq 2p \leq n$ $(n \geq 5)$, $\Delta(G) < n - 1$ *and* $n \leq e(G) \leq n + \frac{n}{p} - 2$. *Then except when* $T = P_2$ *and* $G = O_2 \cup K_4$, T *can be packed into* G *such that* T *covers at least* $p + 2$ *edges of* G *and* $\Delta(G - T) < n - p - 1$.

Proof. The proof is by induction on n and p. The proof that the lemma is true for $p = 2$ is left as an exercise (Ex.4.8(1)).

We now prove that the lemma is true for $p = 3$. Suppose T is a path xyz. We map x to a vertex u of G such that $d(u) = \Delta(G)$, y to a vertex $v \notin N(u)$ and is of maximum possible valency, and z to a vertex w $(\neq u)$ such that $w \notin N(v)$ and is of maximum possible valency. Then T covers at least five edges of G. Now if $\Delta(G - T) \geq n - 4$, then $3(n - 4) \leq \frac{4}{3}n - 2$, from which it follows that $n = 6$ and $e(G) = 6$. However, when $n = 6$ and $e(G) = 6$, the lemma is clearly true.

Hence we suppose that $p \geq 4$ and $n \geq 8$. Since $n \leq e(G) \leq n + \frac{n}{p} - 2$, $\delta(G) \leq 2$. The case that $\delta(G) = 0$ can be settled in a similar way as in the proof of Lemma 4.1. The case that $\delta(G) = 1$ can also be settled in a similar way as in the proof of Lemma 4.1 except that the star-component S_k $(k \geq \frac{n}{2} + 1)$ has to be replaced by an (m, m) graph H such that H has no end-vertex u whose neighbour v has valency $d(v) < \frac{n}{2}$. Finally, suppose $\delta(G) = 2$. Let u be a vertex of G such that $d(u) = 2$

and let $N(u) = \{v,w\}$. The case that $\Delta(G - u) = n - 2$ can be settled as in Case 2 of the proof of Lemma 4.1. Hence we assume that $\Delta(G - u) < n - 2$. We add a suitable edge e to $G - u$ so that $G' = G - u + e$ is such that $\Delta(G') < n - 2$ and $n - 1 \leq e(G') \leq n - 1 + \frac{n}{p} - 2 \leq n - 1 + \frac{n-1}{p-1} - 2$. Thus, by induction, if T is a tree of order $s \leq p - 1$, there is a required packing σ of T and G'. If T covers at least $s + 3$ edges of G', or if T covers exactly $s + 2$ edges of G' which includes the edge e, but $v \in \sigma(T)$ or $w \in \sigma(T)$, then T covers at least $s + 2$ edges of $G - u$ and σ is a required packing of T and G. However, if $v, w \notin \sigma(T)$, we can map an appropriate vertex of T to u so that σ can be modified to yield a required packing of T and G. The remaining case is that $p = \frac{n}{2}$. But in such case, $e(G) = n$ and G is a union of some cycles and the lemma is obviously true. //

We shall now give a complete characterization for packing two graphs G and H of order $n \geq 5$ such that $\Delta(G), \Delta(H) < n - 1$ and $e(G) + e(H) \leq 2n - 2$.

In view of Theorem 7.2, we can assume that G is an $(n, n - k)$ graph and H is an $(n, n + k - 2)$ graph where $k \geq 2$.

We first consider the case $k \geq 3$. In this case, $\Delta(H) \geq 3$. Since $e(G) = n - k$, $k \geq 3$, G has at least k tree-components. Let T be a tree-component of G whose order p is minimum among all the tree-components of G. Then $kp \leq n$ and thus $k \leq n/p$. Hence $e(H) \leq n + \frac{n}{p} - 2$. By Lemma 8.1, there is a packing of T into H so that $H' = H - T$ is an (m, q) graph where $m = n - p$, $q \leq m + k - 4$ and $\Delta(H') < m - 1$. Now if $G' = G - T$ and H' are packable, then G and H are packable. Hence for $k \geq 3$, the forbidden pairs $\{G, H\}$ are generated from the forbidden pairs $\{G', H'\}$ which are given in Theorem 4.2. By examining the forbidden pairs givenin Theorem 4.2, we know that there are no forbidden pairs for $k \geq 5$; for $k = 4$, the forbidden pairs $\{G', H'\}$ are $\{3K_2, O_2 \cup K_4\}$, $\{O_3 \cup K_3, 2K_3\}$ and $\{O_5 \cup K_4, 3K_3\}$; and for $k = 3$, the forbidden pairs $\{G', H'\}$ are $\{2K_2, O_1 \cup K_3\}$, $\{O_2 \cup K_3, K_2 \cup K_3\}$, $\{2K_2 \cup K_3, O_3 \cup K_4\}$ and $\{O_4 \cup K_4, K_2 \cup 2K_3\}$.

It is not difficult to verify that $\{3K_2, O_2 \cup K_4\}$ generates (1) : $\{4K_2, O_3 \cup K_5\}$, $\{O_3 \cup K_3, 2K_3\}$ generates (2): $\{O_4 \cup K_3, K_3 \cup K_4\}$, and

$\{O_5 \cup K_4, 3K_3\}$ generates (3): $\{O_6 \cup K_4, 2K_3 \cup K_4\}$.

It is also not difficult to verify that $\{2K_2, O_1 \cup K_3\}$ generates (4): $\{O_1 \cup 2K_2, O_1 \cup K_4\}$, (5): $\{3K_2, K_2 \cup K_4\}$, (6) and (7); $\{O_2 \cup K_3, K_2 \cup K_3\}$ generates (8): $\{O_3 \cup K_3, K_2 \cup K_4\}$ and (9); $\{2K_2 \cup K_3, O_3 \cup K_4\}$ generates (10): $\{3K_2 \cup K_3, O_4 \cup K_5\}$; and $\{O_4 \cup K_4, K_2 \cup 2K_3\}$ generates (11): $\{O_5 \cup K_4, K_2 \cup K_3 \cup K_4\}$ and (12). (The forbidden pairs (6), (7), (9) and (12) are depicted in Fig. 4.11.)

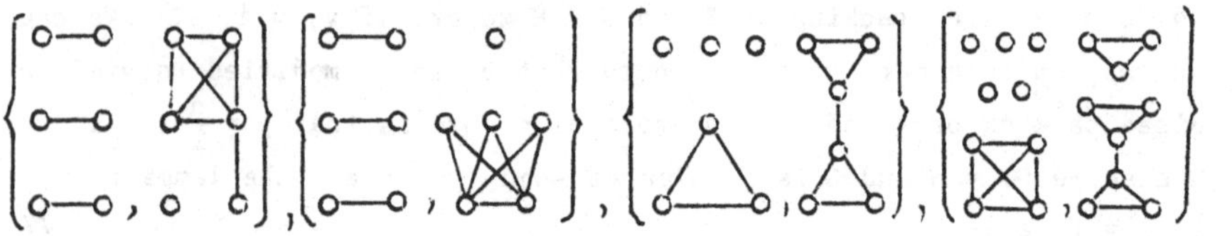

(6) (7) (9) (12): $\{O_5 \cup K_4, H_9\}$

Figure 4.11

The case that $T = P_2$ and $H = O_2 \cup K_4$ yield the forbidden pair (5): $\{3K_2, O_2 \cup K_4\}$.

We next consider the case $k = 2$. First suppose G has an isolated vertex u. Let v be a vertex of H such that $d(v) = \Delta(H)$. Then $G - u$ and $H - v$ are respectively $(n - 1, n - 2)$ and $(n - 1, n - \Delta(H))$ graphs. Again any forbidden pair $\{G, H\}$ must be generated from some forbidden pair $\{G - u, H - v\}$. Now if $\Delta(H) \geq 3$, and $\Delta(G - u) < n - 2$, then by Theorem 4.2, the forbidden pairs $\{G - u, H - v\}$ are $\{O_1 \cup K_3, 2K_2\}$, $\{K_2 \cup K_3, O_2 \cup K_3\}$, $\{O_3 \cup K_4, 2K_2 \cup K_3\}$ and $\{K_2 \cup 2K_3, O_4 \cup K_4\}$. We thus obtain the forbidden pairs (13), (14): $\{O_1 \cup K_2 \cup K_3, O_2 \cup K_4\}$, (15): $\{O_4 \cup K_4, 2K_2 \cup K_4\}$, (16) and (17). (The forbidden pairs (13), (16) and (17) are depicted in Fig.4.12.) On the other hand, suppose

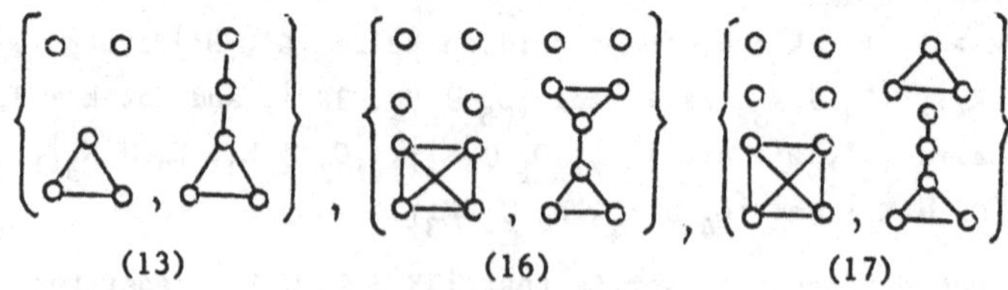

(13) (16) (17)

Figure 4.12

$\Delta(G - u) = n - 2$. Since $\Delta(H) > 3$, H has a vertex w such that $d(w) < 1$. It is easy to see that in this case G and H are packable.

We next consider the case $\Delta(H) = 2$. Since $e(H) = n$, H is the union of some cycles and thus $H - v$ is the union of some cycles and a path of order at least two. Now $G - u$ and $H - v$ are $(n - 1, n - 2)$ graphs. Hence if $G - u \neq S_{n-1}$, $n > 6$, then applying Theorem 7.2, we have the forbidden pairs for $\{G - u, H - v\}$: $\{P_2 \cup K_3, P_2 \cup K_3\}$, $\{G(5), P_2 \cup K_3\}$, $\{O_3 \cup K_4, P_2 \cup C_5\}$, $\{O_3 \cup K_4, P_4 \cup K_3\}$, $\{G(8), P_2 \cup 2K_3\}$, $\{O_2 \cup K_2 \cup K_4, P_2 \cup 2K_3\}$, and $\{O_6 \cup K_5, P_2 \cup 3K_3\}$. From these pairs, we obtain the following forbidden pairs for $\{G, H\}$, namely, (18): $\{O_1 \cup K_2 \cup K_3, 2K_3\}$, (19), (20): $\{O_4 \cup K_4, K_3 \cup C_5\}$, (21), (22): $\{O_3 \cup K_2 \cup K_4, 3K_3\}$ and (23): $\{O_7 \cup K_5, 4K_3\}$. (The forbidden pairs (19) and (21) are depicted in Fig. 4.13.) However if $G - u = S_{n-1}$, then we have the forbidden pair

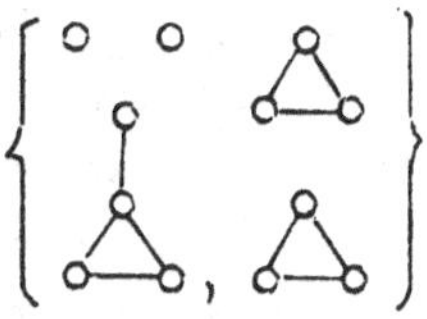

(19)

$\{G(9), 3K_3\} =$

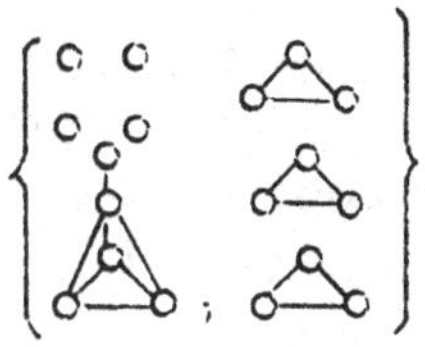

(21)

Figure 4.13

(24): $\{O_1 \cup S_{n-1}, \cup C_i\}$. Also it is clear that there are only two (5,3) graphs $G \neq O_1 \cup S_4$ such that G contains an isolated vertex, namely, $O_1 \cup P_4$ and $O_2 \cup K_3$. In this case we obtain the forbidden pair (25): $\{O_2 \cup K_3, C_5\}$.

Finally we consider the case that G has no isolated vertices. Since G is an $(n, n - 2)$ graph, G has at least two tree-components. Let T be a minimal tree-component of G. Then $2 < |T| = p < \frac{n}{2}$, and by Lemma 8.1, if $H \neq O_2 \cup K_4$, there is a packing of T into H such that T covers at least $p + 2$ edges of H and $\Delta(H - T) < n - p - 1$. Let $G' = G - T$ and $H' = H - T$. Then G' is an $(m, m - 1)$ graph and H' is an (m, q) graph where $q < m - 2$. Thus if $G' \neq S_m$, then the forbidden pairs $\{G, H\}$ are generated from the forbidden pairs $\{G', H'\}$ given in Theorem 4.2 in which G' has a nontrivial tree-component. There are two such pairs, namely, $\{K_2 \cup K_3, O_2 \cup K_3\}$ and $\{K_2 \cup 2K_3, O_4 \cup K_4\}$. From these two pairs, we obtain the forbidden pairs (26): $\{2K_2 \cup K_3, O_1 \cup K_2 \cup K_4\}$, (27) and (28): $\{2K_2 \cup 2K_3, O_5 \cup K_5\}$. (The pair (27) is depicted in Fig.4.14.) On the other hand, if $G' = S_m$, then $G = T \cup S_m$ where

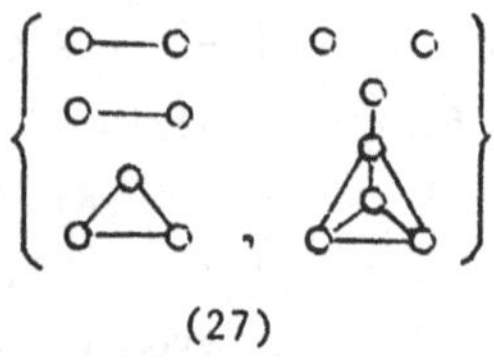

(27)

Figure 4.14

$m > n/2$. In this case, by adding an appropriate edge e to G, we obtain a tree $G + e \neq S'_n$; and by Theorem 6.2, if $n > 7$, there is a packing of $G + e$ and H, unless $H = kC_3$. In this case we obtain the forbidden pair (29): $\{K_2 \cup S_{n-2}, kC_3\}$. For $n < 6$, $G = P_2 \cup S_3$, $P_2 \cup P_4$, $P_2 \cup S_4$ or or $2S_3$. Hence for an appropriate edge e, $G + e = P_5$, S''_6 or P_6. By checking against the forbidden pairs given in Theorem 6.2, we obtain the forbidden pair (30): $\{P_2 \cup P_4, O_2 \cup K_4\}$. It remains to consider the case $H = O_2 \cup K_4$. In this case we obtain the same forbidden pair (30).

The above thirty forbidden pairs, together with the forbidden pairs given in Theorems 4.2 and 7.2, yield all the forbidden pairs for the packing of two graphs G and H of order $n > 5$ such that $\Delta(G), \Delta(H) < n - 1$ and $e(G) + e(H) < 2n - 2$. Because there are too many forbidden pairs, for simplicity, we state the above result only for $n > 9$.

Theorem 8.2 _Suppose_ G _and_ H _are graphs of order_ $n > 9$ _such that_ $\Delta(G), \Delta(H) < n - 1$ _and_ $e(G) + e(H) < 2n - 2$. If $\{G, H\}$ _is not one of the following thirteen pairs:_

$\{3K_3, O_5 \cup K_4\}$, $\{3K_2 \cup K_3, O_4 \cup K_5\}$, $\{O_5 \cup K_4, K_2 \cup K_3 \cup K_4\}$,

$\{O_5 \cup K_4, H_9\}$, $\{G(9), 3K_3\}$, $\{O_3 \cup K_2 \cup K_4, 3K_3\}$, $\{O_6 \cup K_4, 2K_3 \cup K_4\}$,

$\{2K_2 \cup 2K_3, O_5 \cup K_5\}$, $\{O_6 \cup K_5, K_2 \cup 3K_3\}$, $\{O_7 \cup K_5, 4K_3\}$,

$\{O_1 \cup S_{n-1}, \cup C_i\}$, $\{K_2 \cup S_{n-2}, kC_3\}$ _and_ $\{K_3 \cup S_{n-3}, K_3 \cup S_{n-3}\}$,

then G _and_ H _are packable._

Corollary 8.3 Suppose G and H are graphs of order $n \geq 13$. If $\{G, H\}$ is not one of the following three pairs:

$$\{O_1 \cup S_{n-1}, \cup C_i\},\ \{K_2 \cup S_{n-2}, kC_3\} \text{ and } \{K_3 \cup S_{n-3}, K_3 \cup S_{n-3}\},$$

then G and H are packable.

Exercise 4.8

1. Let G be a graph of order $n \geq 5$ such that $\Delta(G) < n - 1$ and $n \leq e(G) \leq \frac{3}{2}n - 2$. Prove that if $G \neq O_2 \cup K_4$, then P_2 can be packed into G such that P_2 covers at least four edges of G and $\Delta(G - P_2) < n - 3$.

2. Suppose $D_i = (V_i, A_i)$, $i = 1, 2$ are digraphs of order n. We say that D_1 and D_2 are packable if there exists a bijection $\sigma : V_1 \to V_2$ such that if $(x, y) \in A_1$ then $(\sigma(x), \sigma(y)) \notin A_2$. Prove that if $|A_1||A_2| < n(n - 1)$, then D_1 and D_2 are packable. Applying the above result, show that if $|A_1| + |A_2| \leq 2n - 2$, then D_1 and D_2 are packable (Benhocine, Veldman and Wojda [83]).

3. Characterize the pairs of digraphs $D_1 = (V_1, A_1)$ and $D_2 = (V_2, A_2)$ of order n, such that $|A_1| + |A_2| = 2n - 1$, for which D_1 and D_2 are not packable (Benhocine, Veldman and Wojda [83]).

4.* For integers k and n satisfying $1 \leq k \leq n(n - 1)$, denote by $f(n,k)$ the minimal number such that there exist digraphs $D_1 = (V_1, A_1)$ and $D_2 = (V_2, A_2)$ of order n, with $|A_1| = k$ and $|A_2| = f(n,k)$, for which there is no packing of D_1 and D_2. Prove that for every m satisfying $2 \leq m \leq \frac{n}{2}$, $f(n, n - m) = 2n - [\frac{n}{m}]$. In particular, $f(n, n - 2) = 2n - [\frac{n}{2}]$ (Wojda [85]).

REFERENCES

A. Benhocine, H. J. Veldman and A. P. Wojda, Packing of digraphs, Technische Hogeschool Twente (Holland) Memorandum No.454 (1983).

B. Bollobás, Extremal Graph Theory, Academic Press, London (1978).

———, Some remarks on packing trees, Discrete Math. 46 (1983), 203-204.

B. Bollobás and S. E. Eldrige, Problem in Proc. Fifth British Combinatorial Conf. (Eds. C. St. J. A. Nash-Williams and J. Sheeham), Utilitas Math., Winnipeg (1976), 689-691.

———, Packing of graphs and applications to computational complexity, J. Combin. Theory, Ser.B, 25 (1978), 105-124.

———, Problem in Proc. Colloque Intern. CNRS (Eds. J. C. Bermond, J. C. Fournier, M. das Vergnas and D. Sotteau), 1978.

D. Burns and S. Schuster, Every (p, p - 2) graph is contained in its complement, J. Graph Theory 1 (1977), 277-279.

———, Embedding (n, n - 1) graphs in their complements, Israel J. Math., vol.30, no.4 (1978), 313-320.

P. A. Catlin, Subgraphs of graphs,I, Discrete Math. 10 (1974), 225-233.

———, Embedding subgraphs under extremal degree conditions, Proc. of the 8th Southeastern Conf. on Combinatorics, Graph Theory and Computing, Baton Rouge, Louisiana (1977), 139-145.

P. Erdös and T. Gallai, On maximal paths and circuits of graphs, Acta Math. Acad. Sci. Hungar. 10 (1959), 337-356.

R. J. Faudree, C. C. Rousseu, R. H. Schelp and S. Schuster, Embedding graphs in their complements, Czechoslovak J. Math. 31:106 (1981), 53-62.

F. Fink and H. J. Straight, A note on path-perfect graphs, Discrete Math. 33 (1981), 95-98.

P. C. Fishburn, Packing graphs with odd and even trees, J. Graph Theory 7 (1983), 369-383.

M. R. Garey and D. S. Johnson, Computer and Intractability, W. H. Freeman, San Francisco (1979).

A. Gyárfàs and J. Lehel, Packing trees of different order into K_n, Colloq. Math. Soc. Jànos 18 (1978), 463-469.

F. Harary, Graph Theory, Addison Wesley, Reading, Massachusetts (1969).

S. M. Hedetniemi, S. T. Hedetniemi and P. J. Slater, A note on packing two trees into K_n, *Ars Combinatoria* 11 (1981), 149-153.

A. M. Hobbs, Packing trees, *Proc. of the 12th Southeastern Conf. on Combinatorics, Graph Theory and Computing*, *Congre. Numer.* 33 (1981), 63-73.

———, Packing sequences, *J. Combin. Theory*, Ser.A, 36 (1984), 174-182.

C. Huang and A. Rosa, Decomposition of complete graphs into trees, *Ars Combinatoria* 5 (1978), 26-63.

E. C. Milner and D. J. A. Welsh, On the computational complexity of graph theoretical properties, *University of Calgary Research Paper* No. 232, June, 1974.

G. Ringel, Problem 25 in *Theory of Graphs and Its Applications*, Proc. Int. Symp. Smolenice 1963, Czech Acad. Sci., Prague, Czech., (1964), p.162.

N. Sauer and J. Spencer, Edge disjoint placement of graphs, *J. Combin. Theory*, Ser.B, 25 (1978), 295-302.

P. J. Slater, S. K. Teo and H. P. Yap, Packing a tree with a graph of the same size, *J. Graph Theory* 9 (1985), 213-216.

H. J. Straight, Packing trees of different size into the complete graph, *Topics in Graph Theory* (Ed., F. Harary), Annals of the New York Academy of Science, vol.328 (1979), 190-192.

S. K. Teo, Packing of Graphs, M. Sc. Thesis, Department of Mathematics, National University of Singapore (1985).

S. K. Teo and H. P. Yap, Two theorems on packing of graphs, *Europ. J. Combinatorics* (to appear).

———, Packing two graphs of order n having total size at most 2n - n (submitted).

A. P. Wojda, Unsolved problems, in *Graph Theory Newsletter*, vol.14, no.2 (March, 1985).

S. Zaks and C. L. Liu, Decomposition of graphs into trees, *Proc. of the 8th Southeastern Conf. on Combinatorics, Graph Theory and Computing*, Baton Rouge, Louisiana (1977), 643-654.

5. COMPUTATIONAL COMPLEXITY OF GRAPH PROPERTIES

1. Introduction and definitions

We can instal a graph G of order n into a computer by encoding the entries of the upper triangular part of its adjacency matrix. One problem arises naturally : "Can we find, in the worst case, whether the graph G has a specific property P, without decoding all the $n(n-1)/2$ entries of the upper triangular part of its adjacency matrix?"

The main objective of this chapter is to introduce a Two Person Game to tackle the above problem.

Let G^n be the set of all graphs of order n and let $F \subseteq G^n$ be the set of all graphs such that each of its members has property P. To see that whether a graph G (of order n) possesses property P or not, it is equivalent of showing whether G belongs to F or not. Hence we can introduce the Two Person Game in a general setting and treat the graph property as a special case.

Let T be a finite set of cardinality $|T| = t$ and let $p(T)$ be the power set of T, i.e. the set of all subsets of T. We call $F \subseteq p(T)$ a property of T. A measure of the minimum amount of information necessary, in the worst case, to determine membership of F is as follows. Suppose two players, called the Constructor (Hider) and Algy (Seeker), play the following game which we also denote by F. Algy asks questions of the Constructor about a hypothetical set $H \subset T$. His questions being of the form "does the element x (of T) belong to H?" to which the Constructor answers "yes" or "no". The Constructor does not need to have any particular set H in mind to begin with but as he answers Algy's questions he is effectively constructing the set H. Indeed, if Algy probes all the elements of T, then he will know precisely which set $H \subset T$ the Constructor is describing. In playing the game F, Algy tries to select questions which enable him to decide as quickly as possible whether or not the set H being constructed by the

Constructor is a member of F. The Constructor on the other hand tries to keep Algy guessing for as long as possible. The (computational) <u>complexity</u> of F, denoted by c(F), is the minimum number of probes needed by Algy to determine membership of F assuming both Algy and the Constructor play the game optimally. If $c(F) = t$, so that Algy has to make all possible probes, then F is said to be <u>elusive</u>, and in this case the Constructor wins the game F. Otherwise F is <u>non-elusive</u> and Algy wins.

In order to systematize our discussion and to avoid possible misunderstanding, we now give a more formal definition of the complexity of a property F. A <u>preset</u> of T is an ordered pair $X = (E,N)$ of disjoint subsets E and N of T. Let PR(T) denote the collection of all presets of T. For $X = (E,N) \in PR(T)$, denote by U(X) the union $E \cup N$. The preset $X = (E,N)$ is <u>proper</u> if $U(X) \neq T$, and <u>full</u> if $U(X) = T$. Let PR*(T) denote the set of all proper presets of T. The preset $Y = (E',N')$ is said to be an <u>extension</u> of $X = (E,N)$ if $Y \neq X$ and $E' \supseteq E$, $N' \supseteq N$, and in this case we write $X < Y$. An <u>algorithm</u> on T is a function $\phi : PR^*(T) \to T$ such that $\phi(X) \notin U(X)$ for any $X \in PR^*(T)$. We call $\phi(X)$ the <u>probe</u> prescribed by the algorithm ϕ for the preset X. A <u>strategy</u> on T is a function $\psi : PR^*(T) \times T \to PR(T)$ such that for $X = (E,N) \in PR^*(T)$ and $x \in T$, $\psi(X,x) = X$ if $x \in U(X)$ and $\psi(X,x) = (E \cup \{x\},N)$ or $(E,N \cup \{x\})$ if $x \notin U(X)$.

Let A(T) and S(T) denote respectively the sets of all algorithms and strategies on T. A pair $(\phi,\psi) \in A(T) \times S(T)$ generates a sequence of presets of T

$$X(\phi,\psi) = \langle X_0, X_1, \ldots, X_t \rangle$$

where $X_0 = (\phi,\phi)$, $X_{i+1} = \psi(X_i, \phi(X_i))$, $i < t$ and X_t is a full preset. Thus $X(\phi,\psi)$ is the sequence of presets given by the Constructor in response to the successive probes $\phi(X_1)$, $\phi(X_2)$, ... by Algy.

Now let F be a property of T. For $A, B \subseteq T$ we write $A \equiv B \bmod F$ if either (i) $A \in F$ and $B \in F$ or (ii) $A \notin F$ and $B \notin F$. Similarly, for presets $X = (E,N)$ and $X' = (E',N')$ we write $X \equiv X' \bmod F$ if and only if $E \equiv E' \bmod F$. A preset X is <u>determining</u> for F, or F-determining, if $X \equiv Y \bmod F$ for every $Y \in PR(T)$ such that $X < Y$. Thus Algy does not need to make further probes once a determining preset for F has been reached.

Next, for $F \subset p(T)$ and $(\phi,\psi) \in A(T) \times S(T)$, we define $c(F,\phi,\psi) = \min \{k \mid X_k \text{ is F-determining}\}$ where $\langle X_0, X_1, \ldots, X_t\rangle = X(\phi,\psi)$. The complexity of F can now be defined as

$$c(F) = \min_{\phi\in A(T)} \max_{\psi\in S(T)} c(F,\phi,\psi).$$

Thus if the property F is elusive, then the Constructor has a winning strategy ψ such that $c(F,\phi,\psi) = t$ for all $\phi \in A(T)$, and if F is non-elusive, then Algy has a winning algorithm ϕ such that $c(F,\phi,\psi) < t$ for all $\psi \in S(T)$.

An element x of T is said to be <u>critical</u> for the preset $X = (E,N)$ in the game F if $x \notin U(X)$, X is not determining but either $(E \cup \{x\}, N)$ or $(E, N \cup \{x\})$ is determining for F. A property F of T is <u>nontrivial</u> if $F \neq \phi$ and $F \neq p(T)$. F is <u>monotone</u> (downwards) if $A \subset B \in F$ implies that $A \in F$. The <u>enumerating polynomial</u> for F is

$$P_F(z) = \sum_{i \leq t} N(F,i)\, z^i$$

where $N(F,i)$ is the number of i-element members in F. The group of permutations on T that leave F invariant is denoted by $\Gamma(F)$. If $X \subseteq T$, $\Gamma(X)$ is the stabilizer of X. Hence, if $X \in F$, $\Gamma(X)$ is a subgroup of $\Gamma(F)$.

Let G^n be the set of all graphs having vertex set $\{0,1,\ldots,n-1\}$. Since a member $G \in G^n$ is uniquely determined by its edge set, we shall not distinguish between the graph G and its edge set E(G). A <u>graph (theoretical) property</u> is a set $F \subseteq G^n$, which is closed under isomorphism, i.e. if $X, Y \in G^n$ and $X \cong Y$ (X and Y are isomorphic), then $X \equiv Y \bmod F$. The <u>capacity</u> of a graph property F is the number of non-isomorphic graphs in F. We call the elements of the edge set $\{ij \mid 0 \leq i < j \leq n-1\}$ of K_n <u>places.</u> Thus if F is a graph property and if Algy and the Constructor play the game F described previously, then Algy will successively probe different places of K_n and the Constructor will indicate whether these probed places are edges or non-edges. We call a preset (E,N) of G^n a <u>pregraph.</u>

In this chapter, we shall find some elusive graph properties and some non-elusive graph properties. We shall study the diagram of a non-elusive property, find a necessary condition (the so-called odd-even balanced condition) for a property to be elusive, discuss the Aanderaa-Rosenberg Conjecture, give a counterexample to the Rivest-Vuillemin Conjecture, and find a lower bound for the computational complexity of general graph properties. Some unsolved problems and conjectures will also be given. The main reference of this chapter is the last chapter of Bollobás' book : Extremal Graph Theory (Academic Press, 1978).

Exercise 5.1

1. Find all the members of the set of Hamiltonian graphs of order 5.

2. Let F be a property of T and let

$$F^* = \{X \subseteq T \mid T - X \notin F\}.$$

Prove that $(F^*)^* = F$ and that F is monotone if and only if F^* is monotone.

3. Show that if F is a nontrivial property of T such that $T \notin F$, then

$$F^{mon} = \{X \mid X \subseteq Z \text{ for some } Z \in F\}$$

is a nontrivial monotone property.

4. Prove that, for any algorithm ϕ, there is an algorithm ϕ' in which Algy always probes a critical element first, if there is one, and is such that the inequality

$$c(F,\phi',\psi) \leq c(F,\phi,\psi)$$

holds for any strategy ψ (Milner and Welsh [74]).

5. Prove that $c(F^*) = c(F)$ for any graph property F (van Emde Boas and Lenstra [74]).

6.* Is it true that $c(F^*) = c(F)$ for any property F?

7. Let F be a property of T and let $F^c = p(T) - F$. Prove that ψ is a winning strategy for F if and only if it is a winning strategy for F^c.

8.* Let F be a property of T and let

$$\bar{F} = \{Y \mid Y = T - X, X \in F\}.$$

Prove or disprove that $c(\bar{F}) = c(F)$.

2. Some elusive properties; the simple strategy ψ_0

Holt and Reingold [72] were the first to prove that the (computational) complexity of certain properties of directed graphs of order n, like being strongly connected or cycle-free, have order $O(n^2)$. Hopcroft and Tarjan [73] and Kirkpatrick [74] also obtained lower bounds on the complexity of certain graph properties. In fact, most of these properties can be proved to be elusive.

Best, van Emde Boas and Lenstra [74] and Milner and Welsh [74] independently introduced a "simple" strategy for the Constructor. In many situations this "simple" strategy is actually winning for the Constructor and using this many natural graph properties like planarity can be shown to be elusive. However, there are situations where the graph property is elusive and yet this "simple" strategy fails, and to find a winning strategy for the Constructor in such cases may be quite difficult. For example, see Bollobás' proof [76] that for $2 \leq r \leq n$ the property "$G \in \mathcal{G}^n$ contains a complete subgraph of order r" is elusive. In this section, we shall first define this simple strategy ψ_0 and use it to prove that several graph properties are elusive. We shall also apply some other strategies to produce further elusive graph properties.

The <u>simple strategy</u> ψ_0 is defined as follows : for any proper preset $X = (E,N)$ and $x \in T - (E \cup N)$, $\psi_0(X,x) = (E \cup \{x\},N)$ if and only if there is $Y \in F$ such that $E \cup \{x\} \subseteq Y$. The following theorem is due to Bollobás [78]. This theorem extends a theorem of Milner and Welsh [74]. It gives a necessary and sufficient condition ensuring that ψ_0 is a winning strategy. We shall use the following lemma.

Lemma 2.1 <u>The simple strategy</u> ψ_0 <u>is a winning strategy in the game</u> F <u>in which</u> $T \notin F$ <u>if and only if it is a winning strategy in the game</u> F^{mon} <u>where</u> $F^{mon} = \{X \mid X \subseteq Z$ <u>for some</u> $Z \in F\}$.

Proof. Suppose ψ_0 is a winning strategy in the game F. Then $c(F,\phi,\psi_0) = t$ for any algorithm ϕ. Let

$$X_0 < X_1 < \dots < X_{t-1} < X_t, \quad X_k = (E_k, N_k)$$

be the sequence of presets given by ϕ and ψ_0 and let x_t be the last unprobed place. Then by the definition of ψ_0 and the fact that $T \notin F$, $E_{t-1} \in F$ and x_t is critical. However, by the definition of F^{mon}, $E_{t-1} \in F^{mon}$ and x_t is also critical for the preset X_{t-1} in the game F^{mon}. Consequently, ψ_0 is also winning in the game F^{mon}.

Conversely, suppose ψ_0 is winning in the game F^{mon}. Suppose the sequence of presets given by ψ_0 and any fixed algorithm is as above. Then $E_{t-1} \in F^{mon}$ and x_t is critical. Again, by the definition of F^{mon} and the fact that E_{t-1} is now a maximal element in F^{mon}, $E_{t-1} \in F$. The last unprobed place x_t is also critical for the preset X_{t-1}. Hence ψ_0 is also winning in the game F. //

Theorem 2.2 (Bollobás [78]) Let F be a nontrivial property of T such that $T \notin F$. Then the simple strategy ψ_0 is a winning strategy if and only if whenever $x \in X \in F$ there are an element $y \in T - X$ and a set $Y \in F$ such that $(X - x) \cup \{y\} \subseteq Y$.

Proof. Sufficiency. Assume that $c(F,\phi,\psi_0) = m < t$ for some algorithm ϕ. Let

$$X_0 < X_1 < \ldots < X_m, \quad X_k = (E_k, N_k)$$

be the sequence of presets given by ϕ and ψ_0. By the definition of ψ_0, there is a set U such that $E_m \subseteq U \in F$. Hence for any V satisfying $E_m \subseteq V \subset T$, $V \in F$ and in particular $X = T - N_m \in F$. Let $x \in T - (E_m \cup N_m)$. Then by the assumption, there are an element $y \in T - X = N_m$ and a set $Y \in F$ such that $(X - x) \cup \{y\} \subset Y$. Now if $k + 1 = \min \{\ell \mid y \in N_\ell\}$, then $(E_k \cup \{y\}, N_k) < (E_m \cup \{y\}, N_m - y)$ and thus $E_m \cup \{y\} \subseteq (X - x) \cup \{y\} = Y \in F$, contradicting the fact that ψ_0 chose y a non-element.

Necessity. Suppose that ψ_0 is a winning strategy in the game F. By Lemma 2.1 and Ex.5.1(3), we can assume, without loss of generality, that F is monotone. Suppose now that there exist $X \in F$ and $x \in X$ such that if $y \in T - X$ then $(X - x) \cup \{y\}$ is not contained in any member of F. Put $s = |X|$ and let ϕ be the algorithm that the first $s - 1$ probes are on the places whose members are the elements of $X - x$ and the last (i.e. the tth) probe is whether x is an element of the hypothetical set or not. Then the $(t-1)^{st}$ preset given by ϕ and ψ_0 is $(X - x, T - X)$. Now there are only two sets extending this preset, namely, $X - x$ and X.

Since both of them belong to F, $c(F,\phi,\psi_0) \leq t - 1$, contradicting the assumption that ψ_0 is a winning strategy. //

Corollary 2.3 The following properties of G^n are elusive:

(1) graphs having at most k edges where $0 \leq k < \binom{n}{2}$;

(2) forests of size k where $1 \leq k < n$.

In the following, we give a nontrivial application of Theorem 2.2 (see Bollobàs [78;p.408]).

Theorem 2.4 (Best, van Emde Boas and Lenstra [74]) For $n \geq 5$, planarity is an elusive property of G^n.

Proof. Suppose F is the set of all planar graphs of order n. We first note that $K_n \notin F$. We shall show that the simple strategy ψ_0 is a winning strategy for the Constructor. Suppose this is not so. Then by Theorem 2.2, there exist a maximal planar graph G and ab ε E(G) such that G is the only maximal planar graph (of order n) containing G - ab. Fix a planar embedding G' of G in the plane. Since G is maximal planar, all the faces of G' are triangles. In particular, ab is the edge of two neighbouring triangular faces, say abc and abd. Then cd ε E(G), for otherwise G - ab + cd would be another maximal planar graph containing G - ab. Let cde and cdf be the two faces containing cd. If {e,f} = {a,b}, then abc, abd, cda, cdb are all the faces of G', so n = 4, contradicting the assumption.

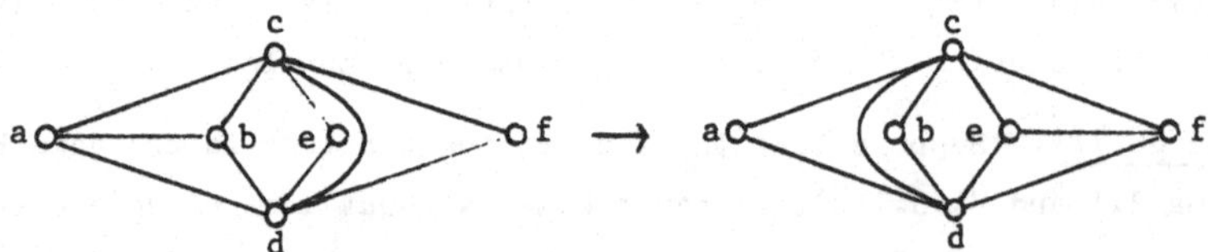

Figure 5.1

Hence we may assume that a ∉ {e,f}. Then ef ∉ E(G) since it would intersect either cd or the path cad. Change the drawing of G - ab as follows : join c to d inside the quadrilateral acbd instead of inside the quadrilateral ecfd. Then we can join e to f inside the quadrilateral ecfd (see Fig.5.1), contradicting the assumption that G is the only planar graph containing G - ab. //

We now give two more examples of elusive graph properties. These examples indicate that very often different strategies have to be used for different graph properties.

Milner and Welsh [74] used an inductive argument to show that for $n \geq 3$ the property of graphs G of order n having a vertex of valency at least 2 is elusive. Their method fails for higher valencies of vertices of G. In the proof of the following example a different approach similar to theirs is used.

Example 1 The property of graphs of order $n \geq 4$ having a vertex of valency at least 3 is elusive.

Proof. Let S be the set of all pregraphs (E,N) such that (i) E has no vertex of valency at least 3, (ii) E does not contain a cycle C_r with $r < n - 1$ and (iii) if E is a path P_n having end-vertices x and y, then $xy \in N$.

We define a strategy ψ_S as follows : for any pregraph X = (E,N) and any probe $x \in T_n - (E \cup N)$, where $T_n = E(K_n)$, put $\psi_S(X,x) = (E \cup \{x\}, N)$ if and only if $E \cup \{x\}$ can be extended to a member of S.

It is not difficult to show that the penultimate pregraph (E_{t-1}, N_{t-1}) in the sequence

$$(\phi,\phi) = (E_o,N_o) < (E_1,N_1) < \dots < (E_{t-1},N_{t-1}) < (E_t,N_t)$$

described by the Constructor using ψ_S is either $E_{t-1} = P_n$ having end-vertices x, y and $xy \in N_{t-1}$ or $E_{t-1} = C_{n-1} \cup P_1$. In either case, it is clear that while E_{t-1} has no vertex of valency at least 3, $E_{t-1} \cup \{e_t\}$, where e_t is the last probe, does have such a vertex. //

The strategy used in proving the next example (Yap [84]) seems to be slightly more complicated.

Example 2 The property of graphs of order $n \geq 2$ which are connected and have a vertex of valency 1 is elusive.

Proof. The Constructor answers "edge" to Algy's first probe $e_1 = xy$. For any pregraph (E,N), the centre of E, denoted by $C(E)$, is the set

$$C(E) = \{x,y\} \cup \{z \mid d_E(z) \geq 2\}.$$

Let S be the set of all pregraphs (E,N) such that (i) $xy \in E$, (ii) for every $u,v \in C(E)$ such that $u \neq v$, $uv \in E \cup N$ and (iii) E has only one nontrivial component, i.e. vertices not connected to the centre are isolated vertices.

Let ψ be the strategy such that for any pregraph $X = (E,N)$ and any probe $x \in T_n - E \cup N$, we put $\psi(X,x) = (E \cup \{x\},N)$ if and only if $E \cup \{x\}$ can be extended to a member of S. We shall show that ψ is a winning strategy.

Let $e_1, e_2, \ldots, e_t$ be any sequence of probes by Algy and let $(\{e_1\},\phi) = (E_1,N_1) < (E_2,N_2) < \ldots < (E_t,N_t)$ be the corresponding sequence of pregraphs given by the Constructor using ψ. We will show that E_{t-1} has the stated property while $E_t \cup \{e_t\}$ does not, so that Algy cannot dispense with his last probe.

First we observe that any $z \notin C(E_{t-1})$ is joined to $C(E_{t-1})$ by an edge. Since at least one of the two places xz, yz is different from e_t, there is a least index $i < t$ such that e_i is a probe between z and the centre $C(E_{i-1})$. However, since $d_{E_{i-1}}(z) = 0$, the strategy ψ gives "e_i is an edge". Hence e_i joins z to a vertex in $C(E_{t-1})$.

Observe that using ψ, the final probe e_t cannot be between two vertices of $C(E_{t-1})$, and since $|C(E_j)| \geq 2$ for any $j \geq 1$, there is one vertex w not in $C(E_{t-1})$. Hence the last probe e_t must be between w and a vertex of $C(E_{t-1})$. Finally, by using ψ, it is clear that E_{t-1} is connected and by the above proof, w is not joined to $C(E_{t-1})$ by an edge. Hence the last probe e_t is critical. //

Exercise 5.2

1. Prove that connectedness is an elusive graph property.
2. Prove that for $n \geq 3$ the property of being a 2-connected graph of order n is elusive.
3.$^-$ Prove that the following properties of G^n, $n \geq 4$, are elusive :

(i) vertex-transitivity,

(ii) edge-transitivity.

4. Let $E \subset G^n$ be the set of connected graphs having an Eulerian trail. Prove that E is elusive (Yap [84]).

5.[+] Let $2 \leq r \leq n$. Prove that the property of containing a complete subgraph of order r is elusive (Bollobás [76]). (Note that Bollobás' proof also shows that the property $\chi(G) \geq r$ is elusive for $2 \leq r \leq n$.)

6. Let n be a prime. Prove that the property being Hamiltonian graphs of order n is elusive (Best, van Emde Boas and Lenstra [74]).

7. Prove that the property of graphs of order $n \geq 3$ having a vertex of valency at least 2 is elusive (Milner and Welsh [74]).

3. Some non-elusive properties

Comparatively it is easier to find elusive graph properties than to find non-elusive graph properties. Up to now only a few non-elusive graph properties have been found. Among the existing non-elusive graph properties, the property of being a scorpion graph (see Fig.5.2 and the definition given below) of order n whose complexity is at most 6n is the most interesting one. This result was proved by Best, van Emde Boas and Lenstra [74] (see also Bollobás [78;p410]).

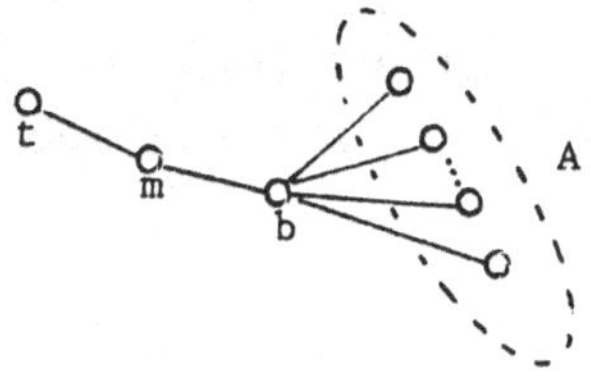

a scorpion graph

Figure 5.2

(A scorpion graph of order $n \geq 6$ is a graph G having a body vertex b of valency $n - 2$ and a tail vertex t of valency 1 which is adjacent to a vertex m of valency 2, and all the other $n - 3$ vertices A of G may or may not be adjacent as shown in Figure 5.2.)

Theorem 3.1 *If* F *is the property of being a scorpion graph of order* n, *then* $c(F) < 6n$.

Proof. We shall find a winning algorithm for Algy. Since $c(F) < \binom{n}{2}$, we may assume that $n > 14$. Algy's aim is to locate the body vertex b and the tail vertex t of the scorpion graph. Define the *weight* of a body candidate (resp. tail candidate) x as two minus the number of probed places incident with x which have been answered "non-edges" (resp. "edges"). Thus each candidate has weight 2 or 1.

Let $V = V(G) = \{1,2,\ldots,n\}$. First Algy probes the places 12, 23, ..., $(n-1)n$, $n1$. By these probes, V is partitioned into three parts : B_2, the set of body candidates of weight 2 (these cannot be tail candidates); T_2, the set of tail candidates of weight 2 (these cannot be body candidates); and $R = V - B_2 - T_2$.

Algy now probes at most $r = |R|$ places each of which is incident with at least one vertex in R. In so doing, R is partitioned into two parts : B_1, the set of body candidates of weight 1; and T_1, the set of tail candidates of weight 1. It is clear that $B_1 \cap T_1 = \phi$.

At this stage of the game, after at most $n + r$ probes, the set of body candidates $B = B_1 \cup B_2$ is disjoint from the set of tail candidates $T = T_1 \cup T_2$, and the sum of the weights is at most $2n - r$. Algy next probes the places between the body candidates and the tail candidates. Since each probe reduces the total weight by exactly one, this part of the game needs at most $2n - r - 2$ probes. Denote by B' (resp. T') the set of the remaining body candidates (resp. tail candidates), and let $b = |B'|$, $t = |T'|$. If $b, t < 1$, we are at home.

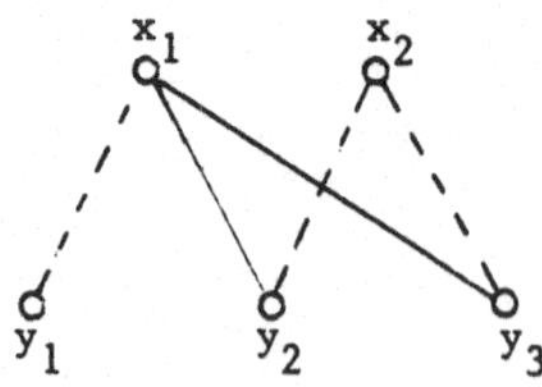

Figure 5.3

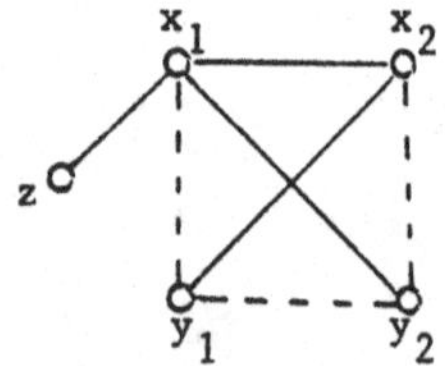

Figure 5.4

Suppose b $\geq$ 2 and t $\geq$ 3. Let $B' = \{x_1, x_2, \ldots\}$ and $T' = \{y_1, y_2, \ldots\}$. If x_1y_1 is a non-edge, then x_1y_2, x_1y_3 must be edges and thus x_2y_2, x_2y_3 are non-edges, which contradicts the fact that x_2 is a body candidate (see Fig.5.3). If x_1y_1 is an edge, then x_2y_1 is a non-edge and we have the previous situation.

Similarly, we can dispose of the case that t $\geq$ 2 and b $\geq$ 3. Hence we need only to consider the following three remaining cases :

Case 1. $B' = \{x_1\}$, $T' = \{y_1,y_2\}$.

Since y_1 and y_2 are tail candidates and x_1 is a body candidate, x_1y_1 and x_1y_2 must be non-edges, which yields a contradiction.

Case 2. $B' = \{x_1,x_2\}$, $T' = \{y_1\}$.

In this case we may assume that x_1y_1 and x_2y_1 are non-edges, otherwise the pregraph cannot be extended to a scorpion graph. Now Algy probes the remaining unprobed places incident with y_1 so that he can locate the vertex m which is adjacent to both the body vertex and the tail vertex. This number is at most n - 3. After that Algy probes all the unprobed places incident with the candidate m so that he can really locate this vertex. This number is at most n - 2. At this stage, the unique body candidate has been found and Algy needs only to probe at most another n - 3 places to decide whether the pregraph can be extended to a scorpion graph or not. Hence the total number of probes required by Algy is at most $(n + r) + (2n - r - 2) + 2(n - 3) + (n - 2) < 6n$.

Case 3. $B' = \{x_1,x_2\}$, $T' = \{y_1,y_2\}$.

In this case we may suppose that x_1y_1, x_2y_2 are non-edges, x_1x_2 is an edge, and y_1y_2 is a non-edge. Otherwise the pregraph cannot be extended to a scorpion graph. Algy now probes a place x_1z for some $z \in V - (B' \cup T')$. If x_1z is an edge, then y_2 cannot be the tail vertex and x_2 cannot be the body vertex of a scorpion graph (see Fig.5.4). If x_1z is a non-edge, then x_1 cannot be the body vertex and y_1 cannot be the tail vertex of a scorpion graph. This again shows that after at most $(n + r) + (2n - r - 2) + 3 = 3n + 1$ probes we can find the unique body candidate and the unique tail candidate.

Finally, it is clear that Algy needs at most another $3(n - 1) - 3$ probes to decide whether the pregraph can be extended to a scorpion graph or not. //

The above theorem shows that the computational complexity of being scorpion graphs of order $n \geq 6$ is bounded by a linear function, and that for $n \geq 14$, this property is non-elusive. In fact, for all $n \geq 6$, this property is non-elusive (see Ex.5.3(7)).

Exercise 5.3

1. Let $n = 2k$, $k \geq 3$. Let $F \subset G^n$ be the set of all graphs G such that G has two adjacent vertices x and y, and if $X = N(x) - y$, $Y = N(y) - x$, then $X \cap Y = \phi$, $|X| = |Y| = k - 1$ and G has no edge joining a vertex in X with a vertex in Y. Give an algorithm proving that

$$c(F) \leq \frac{3n^2}{8} + \frac{n}{4} - 1.$$

E.g. For k = 3,

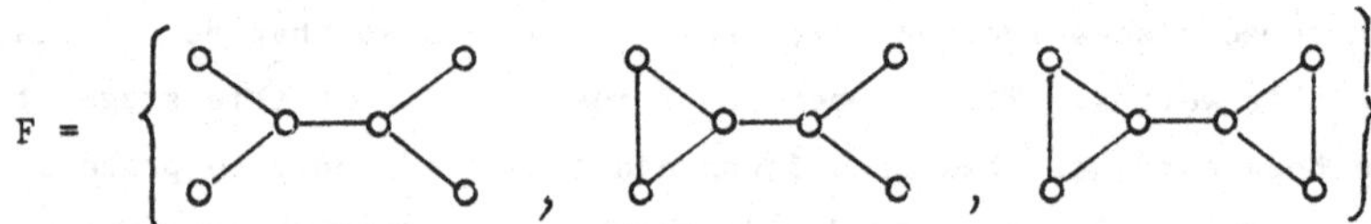

Figure 5.5

(This result is due to D. J. Kleitman. See Best, Van Emde Boas and Lenstra [74].)

2. Prove that the following property of G^7 is not elusive (Milner and Welsh [74]).

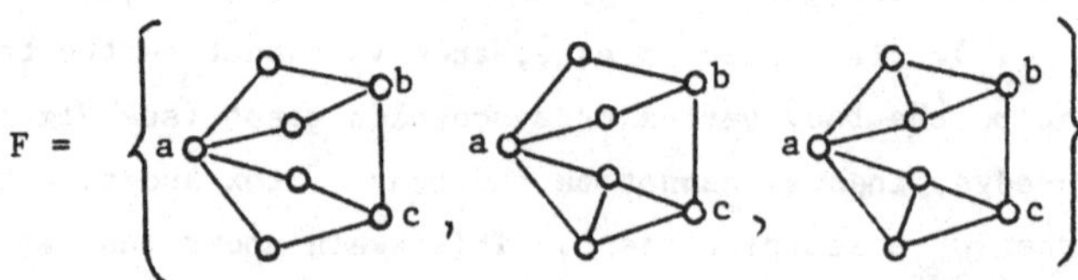

Figure 5.6

3. Define a property F_3 of G^n, $n \geq 7$, as follows: Replace the edge bc in each of the three graphs in Fig.5.4 by a path of length $n - 6$ and let $F_3 \subset G^n$ consist of all graphs isomorphic to one of these three graphs. Prove that F_3 is not elusive (Bollobás [78]).

4. Let F_3' be the following property of G^n, $n \geq 7$. A graph $G \in F_3'$ if and only if G has three vertices a, b, c such that $N(b) \cap N(c) = \phi$ and $N(b) \cup N(c) = V(G) - \{a\}$. Prove that F_3' is not elusive (Milner and Welsh [74]).

5. Let F be the following property of G^n, $n \geq 9$. A graph $G \in F$ has a vertex of valency $n - 4$ and each of the vertices adjacent to this vertex has valency 1. Prove that F is not elusive.

   E.g. For $n = 9$, F =

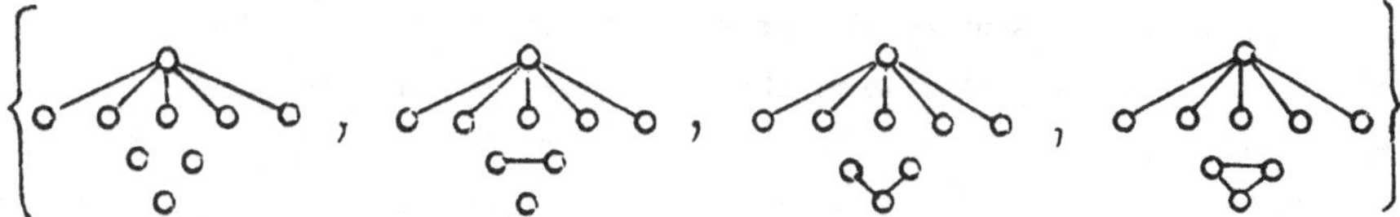

Figure 5.7

   (This result is due to L. Carter. See Best, van Emde Boas and Lenstra [74].)

6. Let F be the property of G^6 given in Fig.5.5 and let $F_c = \{\bar{X} \mid X \in F\}$ where $\bar{X}$ is the complement of X. Prove that $F \cup F_c$ is also not elusive (Yap [84]).

7. Prove that the property of being a scorpion graph of order $n \geq 6$ is non-elusive (Yap [84]).

4. The diagram of a non-elusive property

We recall that if F is a graph property, then the capacity of F is the number of non-isomorphic graphs in F. It is obvious that every property having capacity 1 is elusive. For $n \geq 7$, Milner and Welsh [74] constructed a non-elusive property of G^n having capacity 3 (see Ex.5.3(2)) and they conjectured that every graph property of capacity 2 is elusive. This conjecture was proved by Bollobás and Eldridge [78].

In this section, we shall introduce the notion of the diagram of a property and we shall modify their proof to find some necessary conditions for non-elusive properties having small capacity.

Let F be a property. The diagram of F is a diagraph whose vertices are the elements of F and whose arcs are the ordered pairs (X,Y) where X, Y ϵ F and Y = X $\cup$ {e} for some e ϵ T - X. We denote the diagram of F by D(F). Note that the diagram of a graph property is triangle-free.

Theorem 4.1 The diagram of a non-elusive graph property does not contain an isolated vertex.

Proof. Suppose X ϵ F is an isolated vertex in D(F). For any place x probed by Algy, the Constructor fixes an isomorphic copy G $\cong$ X and answers "x is an element of the preset" if and only if x ϵ G. It is clear that the Constructor wins this game, which contradicts the hypothesis. //

Theorem 4.2 (Yap [84]) Suppose F is a non-elusive property of G^n and Y ϵ F. If either one of the following holds

(i) there is a unique edge e ϵ Y such that Y - e ϵ F;

(ii) there are a unique X ϵ F and distinct edges e, e' $\notin$ X such that

$$X \cup \{e\} \cong Y \cong X \cup \{e'\},$$

then there is Z ϵ F such that (Y, Z) is an arc in D(F).

The dual statement is also true.

Proof. We prove this theorem by contradiction.

Suppose (i) holds and F has no element Z such that (Y,Z) is an arc in D(F). Consider the following strategy ψ_1 by the Constructor. To the first probe e_1 by Algy, the Constructor answers "e_1 is an edge". He then fixes an isomorphic copy Y_1 of Y in which e_1 plays the role of e. For any subsequent probe x by Algy, the Constructor answers "x is an edge" if and only if x ϵ Y_1. Now let $e_1,\ldots,e_t$ ($t = \binom{n}{2}$) be the sequence of probes by Algy and let $(\{e_1\},\phi) = (E_1,N_1) < (E_2,N_2) < \ldots < (E_t,N_t)$ be the corresponding sequence of pregraphs described by the Constructor using the strategy ψ_1. Then either $E_{t-1} = Y_1$ or $E_{t-1} \cup \{e_t\}$

$= Y_1$. However, if $E_{t-1} = Y_1$, then e_t is critical because there is no $f \notin Y_1 \cong Y$ such that $Y_1 \cup \{f\} \in F$. On the other hand, if $E_{t-1} \cup \{e_t\} = Y_1$, then e_t is also critical otherwise $Y_1 - e_t \in F$, contradicting the uniqueness of e.

Suppose (ii) holds and F has no element Z such that (Y,Z) is an arc in D(F). Consider the following strategy ψ_2 by the Constructor in which for any pregraph (E,N) and any probe $x \notin E \cup N$ he answers "x is an edge" if and only if $E \cup \{x\}$ can be extended to an isomorphic copy of Y. Again, let $e_1,\ldots,e_t$ be the sequence of probes by Algy and let $(E_1,N_1) < \ldots < (E_t,N_t)$ be the corresponding pregraphs described by the Constructor using the strategy ψ_2. Then again either $E_{t-1} \cup \{e_t\} \cong Y \in F$ or $E_{t-1} \cong Y \in F$. However, if $E_{t-1} \cong Y$, then e_t is critical, which contradicts the hypothesis. On the other hand, if $E_{t-1} \cup \{e_t\} \cong Y$, then $E_{t-1} \in F$ otherwise F is elusive. Now $E_{t-1} \in F$ implies that $E_{t-1} \cong X$ by the uniqueness of X. If e_t is a unique edge such that $X \cup \{e_t\} \cong Y$, then by (i) there is $Z \in F$ such that (Y,Z) is an arc in D(F), which is a contradiction. Hence we can assume that e_t plays the role of e', say. But then the constructor must have already chosen e an edge and E_{t-1} should in fact be isomorphic to Y, which yields another contradiction.

The dual statement follows from the fact that $F \subset G^n$ is elusive if and only if $G^n - F$ is elusive. //

A <u>star sink</u> in a digraph D = (V,A) is a vertex x such that $(y,x) \notin A$ for any $y(\neq x) \in V$ and if $z \in V$ is such that $(x,z) \in A$, then z is of in-degree 0 and out-degree 1. A <u>star source</u> is similarly defined. The following theorem is due to E. C. Milner (see Yap [84]).

Theorem 4.3 <u>If the diagram of a graph property</u> F <u>has a star source or a star sink, then</u> F <u>is elusive.</u>

Proof. Suppose X is a star sink in D(F) and that the statement is false. (The proof for a star source is similar.) By Theorems 4.1 and 4.2, for each $Y_i(\neq X) \in F$, there is a unique $f_i \notin X$ such that $X \cup \{f_i\} \cong Y_i$ and that there is $f_i' \in X$ so that $(X - f_i') \cup \{f_i\} \cong X$. We now consider the following strategy ψ_3 by the Constructor. For any pregraph (E,N) and any probe $x \notin E \cup N$, he answers "x is an edge" if and only if there is no extension (E',N') of $(E,N \cup \{x\})$ with $E' \cong X$. Let $e_1,\ldots,e_t$ be the sequence of probes by Algy and let $(E_1,N_1) < \ldots < (E_t,N_t)$ be the corresponding sequence of pregraphs described by the Constructor using ψ_3. Since F is assumed non-elusive, it is clear that $E_{t-1} \cong X$. Now since e_t is not critical we must have $E_t = X \cup \{e_t\} \in F$. Thus $e_t = f_i$ for some i. But when Algy probed the place f_i', the Constructor would have answered "f_i' is a non-edge" and this contradicts the fact that $f_i' \in X$. //

From Theorems 4.1 and 4.3, we have

Corollary 4.4 (Bollobás and Edridge [78]) _If_ F _is a graph property of capacity at most_ 2, _then_ F _is elusive._

Corollary 4.5 _If_ $F = \{X_1,X_2,X_3\}$ _is a non-elusive graph property of capacity_ 3, _where_ $|X_1| < |X_2| < |X_3|$, _then_

$$D(F) = \underset{X_1}{\circ} \longleftarrow \underset{X_2}{\circ} \longleftarrow \underset{X_3}{\circ}$$

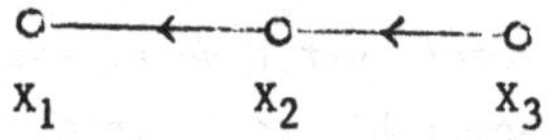

From the proofs of Theorems 4.1, 4.2 and 4.3, we also have

Theorem 4.6 _Let_ F _be a graph property._ _If_ D(F) _has a component which is either an isolated vertex, has a star source, or has a star sink, then_ F _is elusive._

Exercise 5.4

1. Prove that if $F = \{X_1,X_2,X_3,X_4\}$ is a non-elusive graph property of capacity 4, where $|X_1| < |X_2| < |X_3| < |X_4|$, then D(F) is one of the following digraphs

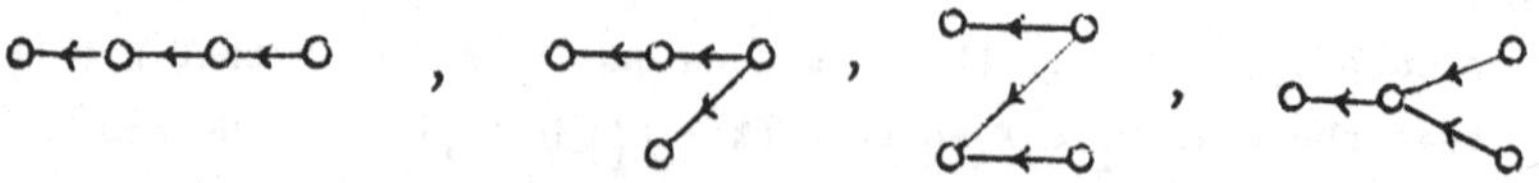

(a) , (b) , (c) , (d)

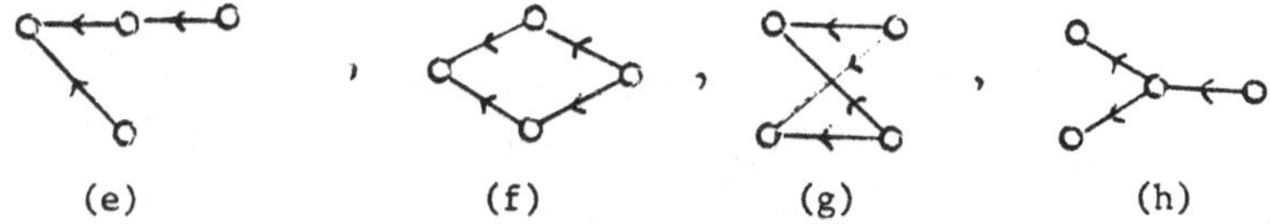

(e) (f) (g) (h)

Figure 5.8

2.* There exist non-elusive graph properties of capacity 4 whose diagram is a dipath of length 4. S. K. Teo (unpublished) has constructed a non-elusive graph property whose diagram is given in Fig. 5.8 (f). Does there exist a non-elusive graph property of capacity 4 whose diagram is one of the other digraphs given in Fig.5.8 ?

3.* Let $F_1,\ldots,F_k \subset G^n$ be non-elusive properties. Suppose $F_i \cap F_j = \phi$ for any $i \neq j$. Is it true that $c(F) = \min\{c(F_i) \mid i = 1,\ldots,k\}$ where $F = F_1 \cup \ldots \cup F_k$?

5. The odd-even balanced condition

A necessary condition for a property F to be elusive was independently found by Rivest and Vuillemin [76] and Best, van Emde Boas and Lenstra [74]. We now give a proof of this result.

Theorem 5.1 (Best, van Emde Boas and Lenstra [74]; Rivest and Vuillemin [76]) Let F be a property of T, $|T| = t$. If $c(F) = k$, then $(1 + z)^{t-k}$ divides the enumerating polynomial $P_F(z)$ of F.

Proof. We need only to prove this theorem for $k < t$. Let ϕ be a winning algorithm. The decision-tree T_ϕ of ϕ is a binary tree whose vertices are labelled with $x_i \in T$. It is rooted at x_1 where x_1 is the first probe by Algy. Suppose the hypothetical set is H. If the Constructor gives $x_1 \notin H$, then the algorithm continues with the left subtree rooted at x_2 where x_2 is the second probe by Algy, and we draw a dotted edge joining x_1 with x_2; otherwise the algorithm continues with the right subtree rooted at x_2' where x_2' is the second probe by Algy, and we draw a solid edge joining x_1 with x_2'. The decision-tree grows at the new root x_2 or x_2', and continues growing in this way. A diagram of a decision-tree is shown below.

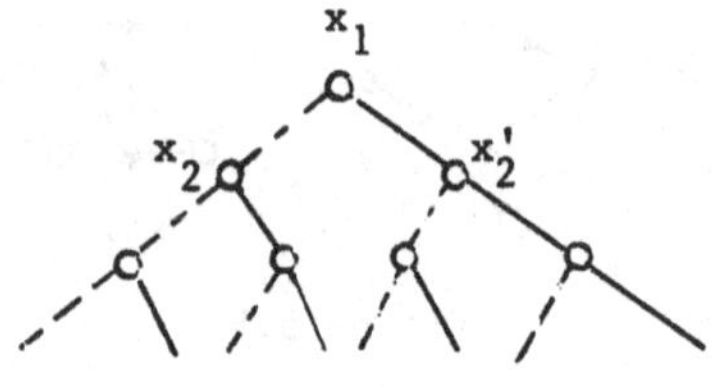

a decision-tree

Figure 5.9

Since $c(F,\phi,\psi) \leq k$, the <u>leaf</u> (i.e. the determining preset), which is eventually reached, specifies whether the preset S belongs to F or not. If it belongs to F, a value 1 is given to it, otherwise a value 0 is given to it. The <u>depth</u> of the leaf is the number of probes asked.

Suppose a leaf is at depth j ($j \leq k$). Suppose the sequence of presets given by ψ is $X_1 < X_2 < \ldots < X_j < \ldots < X_t$, $X_i = (E_i, N_i)$. Suppose X_j is determining and its value is 1. Then $X_j < S = (E,N)$ implies that $E \in F$. Moreover, there is no E_ℓ with $\ell < j$ such that $|E_\ell| < |E_j|$ and $E_\ell \in F$. Thus the contribution of a leaf, at depth j, to $P_F(z)$ is $z(1 + z)^{t-j}$ where $i = |E_j|$. Finally, each $X \in F$ is contained in exactly one branch of the decision-tree with a leaf receiving the value 1, therefore $(1 + z)^{t-k} \mid P_F(z)$. //

Corollary 5.2 (The odd-even balanced condition) <u>If the number of odd-sized elements in</u> F <u>and the number of even-sized elements in</u> F <u>are not equal</u>, <u>then</u> F <u>is elusive</u>.

Proof. If F is elusive and $c(F) = k < t$, then $(1 + z)^{t-k} \mid P_F(z)$, and therefore $P_F(-1) = 0$, a contradiction to the assumption. //

Suppose Γ is a group of permutations of T and $X \subset T$. Then the orbit of X under the action of Γ is the set

$$\mathrm{Orb}_\Gamma(X) = \{\sigma(X) \mid \sigma \in \Gamma\}.$$

Thus if $X_1, \ldots, X_r$ are the distinct members of $\mathrm{Orb}_\Gamma(X)$, then

$$r = |Orb_\Gamma(X)| = |\Gamma|/|\Gamma(X)| \tag{1}$$

In particular, if X is a graph of order n, Γ is the full permutation group on the vertex set $\{0,1,\ldots,n-1\}$, then $\Gamma(X)$ = Aut X is the automorphism group of X. Thus

$$|Orb(X)| = n!/|\text{Aut } X| \tag{2}$$

is the number of isomorphic copies of X.

The following example is an application of Corollary 5.2.

Example. Let

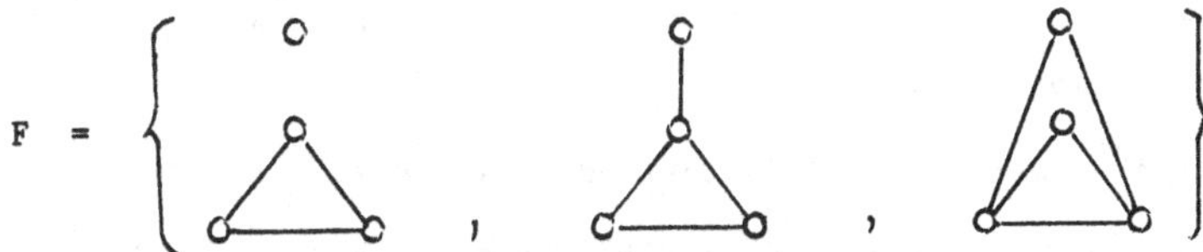

be a property of G^4. Then the number of odd-sized elements in F is 10 and the number of even-sized elements is F is 12. Hence F is elusive.

The following are further applications of Corollary 5.2. These results are due to Rivest and Vuillemin [76]. The proofs of Theorems 5.4 and 5.5 are taken from Bollobás [78].

Theorem 5.3 _Let_ $t = |T| = p^r$, _where_ p _is a prime_. _If_ $\Gamma(F)$ _is transitive on_ T, $\phi \in F$ _and_ $T \notin F$, _then_ F _is elusive_.

Proof. Let $X \in F$. Since $\Gamma = \Gamma(F)$ is transitive on T, each element of T is contained in exactly c of the sets in $Orb_\Gamma(X)$. Hence

$$|Orb_\Gamma(X)||X| = p^r c \tag{3}$$

Thus either $|X| = p^r$ or 0, or else p divides $|Orb_\Gamma(X)|$ which is the number of distinct elements in F isomorphic with X. However since $T \notin F$, $|X| \neq p^r$. Hence the number of distinct elements in F isomorphic with $X \neq \phi$ is always a multiple of p. This shows that the odd-even condition for F is not balanced. Thus F is elusive. //

Theorem 5.4 If F is a nontrivial monotone property of graphs of order $n = 2^m$, then $c(F) \geq n^2/4$.

Proof. Let $H_i = 2^{m-i}K_{2^i}$ be the disjoint union of 2^{m-i} copies of K_{2^i}. Then

$$H_0 = \phi \subset H_1 = \frac{n}{2} K_2 \subset \ldots \subset H_m = K_n.$$

Since F is monotone, there is an index j such that $H = H_j \in F$ and $H_{j+1} \notin F$. Put $J = 2^{m-j-1}K_{2^j}$. Then $H = 2J$ and $H_{j+1} \subset K = J + J$, the join of two copies of J. Hence $K \notin F$.

Being generous to Algy, the Constructor gives away that the hypothetical graph G satisfies $H \subseteq G \subset K$. However, when answering question about places in $T = E(K) - E(H)$, the Constructor will try to play as well as possible.

A property P of T is defined as follows:

$$P = \{E(G) - E(H) \mid H \subseteq G \subset K \text{ and } G \in F\}.$$

Then $\phi \in P$, $T \notin P$ and $c(P) \leq c(F)$. In fact for given G satisfying $H \subseteq G \subset K$, Algy needs at most $c(P)$ probes to decide whether $G \in F$ or $G \notin F$.

It is clear that T and P satisfy the conditions of Theorem 5.3. (If $e, f \in E(K) - E(H)$, then there is a permutation of $V(G)$ mapping H and K into themselves, that maps e to f. Therefore $\Gamma(P)$ is transitive on T.) Now $|T| = n^2/4 = 2^{2m-2}$ is a prime power and thus $c(F) \geq c(P) \geq n^2/4$. //

Theorem 5.5 If F is a nontrivial monotone property of G^n, then $c(F) \geq n^2/16$.

Proof. Let $c(n) = \min \{c(F) : F \subseteq G^n \text{ is nontrivial monotone}\}$. The assertion is an immediate consequence of Theorem 5.4 if we prove the following inequality : if $2^m < n < 2^{m+1}$, then

$$c(n) \geq \min \{c(n-1),\ 2^{2m-2}\} \tag{4}$$

To prove (4), consider a monotone property $F \subset G^n$. If $O_1 \cup K_{n-1} \notin F$ or $S_n \in F$ then, as in the proof of Theorem 5.4, if the Constructor gives away that a certain vertex has degree 0 or $n - 1$, Algy still needs at least $c(n-1)$ probes. Hence we may assume that $O_1 \cup K_{n-1} \in F$ and $S_n \notin F$.

Let $r = 2^{m-1}$ and $s = n - 2r$. Then the monotonicity of F implies that $O_r \cup (K_r + K_s) \in F$, since it is a subgraph of $O_1 \cup K_{n-1}$. Similarly, $K_r + (O_r \cup K_s) \notin F$ since it contains S_n. As in the proof of Theorem 5.4, F can be used to define a transitive property P on the set T of edges joining K_r with O_r and $c(P) \leq c(F)$. The property P satisfies the conditions of Theorem 5.4, so $c(P) \geq r^2$, completing the proof of (4). //

Kleitman and Kwiatowski [80] improved the bound given in Theorem 5.5 to $n^2/9$. Since their proof is complicated and there is still a big gap between this bound and the conjectured value $c(F) = \binom{n}{2}$ (see §6), we shall refer the interested readers to their original paper.

Exercise 5.5

1. Applying the odd-even balanced-condition, prove that every property $F \subset G^n$, $n \leq 4$, is elusive (Milner and Welsh [74]; Yap [84]).

2.* Prove that every property of $F \subset G^5$ is elusive.

3. Applying the odd-even balanced condition, prove that for $n \geq 3$,

$$F = \{O_n, O_{n-2} \cup S_2, \ldots, O_1 \cup S_{n-1}, S_n\}$$

is an elusive graph property (Best, van Emde Boas and Lenstra [74]).

4. Let $F = \{G \in G^n \mid G \text{ contains two non-incident edges}\}$, $n \geq 4$. Applying the odd-even balanced condition, prove that F is elusive.

5. Let F be the property of being a scorpion graph of order 6. Show that $c(F) \geq \binom{6}{2} - 2$.

6. Show that if the capacity of $F \subseteq G^n$ is 3 and that if F is non-elusive, then $c(F) = \binom{n}{2} - 1$ (Yap [84]).

7. Let $t = |T|$. Prove that as $t \to \infty$, almost all the properties of T are elusive (Rivest and Vuillemin [76]).

6. The Aanderaa-Rosenberg Conjecture

Aanderaa and Rosenberg [73] conjectured that there exists a positive number c such that for every nontrivial monotone property F of graphs (which may have loops) of order n, $c(F) > cn^2$. Counter-examples to this conjecture were produced by Lipton and Snyder [74]. These counter-examples all involve with loops and thus the conjecture was modified by Lipton and Synder accordingly. This modified conjecture was subsequently proved by Rivest and Vuillemin [74] (see Theorem 5.5). In order to revive the conjecture, Rivest and Vuillemin changed the conjecture to the following form :

A-R CONJECTURE Every nontrivial monotone graph property is elusive.

In this section, we shall prove that the A-R Conjecture is true for most of the cases. This result is given by Corollary 6.4. The A-R Conjecture has also been proved for all nontrivial monotone properties $F \subset G^n$ where n is a prime power in a very recent paper by Kahn, Saks and Startevant [84]. Their topological approach is nice but difficult to fit in here. We refer the interested readers to their original paper.

Lemma 6.1 Suppose p is a prime divisor of $|T|$. If the group Γ of permutations on T acts transitively on T, then $p \mid |\Gamma|$ and for any $X \subseteq T$,

$$p \nmid |X| \quad \text{or} \quad p \nmid |\Gamma(X)| \quad \text{implies that} \quad p \mid |Orb_\Gamma(X)| \tag{5}$$

Proof. Since $Orb_\Gamma(\{x\}) = T$ for any $x \in T$, it follows from (1) that $p \mid |\Gamma|$. Now for any $X \subseteq T$, we have, by the proof of Theorem 5.3, $p \mid |Orb_\Gamma(X)|\,|X|$. Hence if $p \nmid |X|$, then $p \mid |Orb_\Gamma(X)|$. Also, if $p \nmid |\Gamma(X)|$, then $p \mid |\Gamma|$ and $|\Gamma| = |Orb_\Gamma(X)|$ imply that $p \mid |Orb_\Gamma(X)|$. //

Corollary 6.2 Suppose p is a prime divisor of $\binom{n}{2}$. If $X \in G^n$ and if $p \nmid |X|$ or $p \nmid |Aut\ X|$, then $p \mid n!/|Aut\ X|$.

Proof. This follows from Lemma 6.1 and the equality given by (2). //

The following results are due to Yap [84].

Theorem 6.3 Let F be a property of T such that $\phi \in F$ and $T \notin F$. Suppose p is a prime divisor of $|T|$. If $\Gamma = \Gamma(F)$ acts transitively on T and if $p \nmid |X|$ or $p \nmid |\Gamma(X)|$ for each $X \in F - \{\phi\}$, then F is elusive.

Proof. By Lemma 6.1, if $X \neq \phi$, then $p \mid |Orb_\Gamma(X)|$. Since $|Orb_\Gamma(\phi)| = 1$ is not a multiple of p, the odd-even condition is not balanced. Hence F is elusive. //

Corollary 6.4 Let p be a prime divisor of $\binom{n}{2}$. If F is a nontrivial monotone property of G^n such that for each $X(\neq 0_n) \in F$, either $p \nmid |X|$ or $p \nmid |\text{Aut } X|$, then F is elusive.

The above corollary shows that the A-R Conjecture is true for most of the cases. For instance, let $F \subseteq G^6$ consist of all subgraphs of the following two graphs. Then either $5 \nmid |X|$ or $5 \nmid |\text{Aut } X|$, therefore F is elusive.

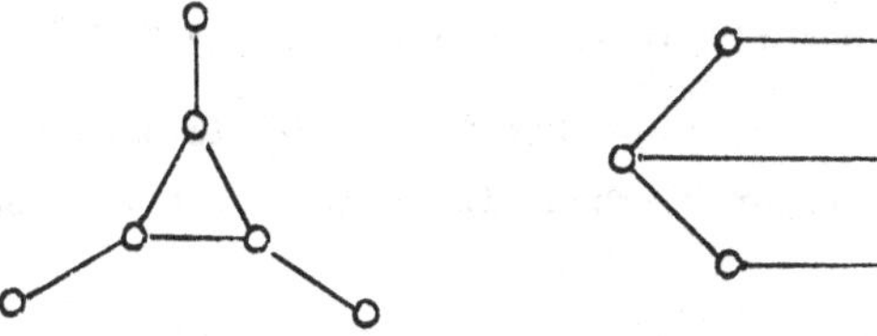

Figure 5.10

Let F be a nontrivial monotone property of G^n and let $p\ (> n)$ be the smallest prime such that $p > \max\{|X| \mid X \in F\}$. For each $X \in F$, we define $X' = X \cup 0_{p-n}$ and $F' = \{X' \mid X \in F\}$. Then F' is a nontrivial monotone property of G^p and for each $X'(\neq 0_p) \in F'$, $p \nmid |X'|$. Thus, by Corollary 6.4, F' is elusive. Now to detect the first isolated vertex in a hypothetical graph H, Algy needs to make at least $p - 1$ probes and to detect the second isolated vertex in H, Algy needs to make at least another $p - 2$ probes and so on. Hence if we can prove that

$$c(F) \geq c(F') - [(p - 1) + \dots + n] \tag{6}$$

then $\binom{n}{2} > c(F) > \binom{p}{2} - [(p-1) + \dots + n] = \binom{n}{2}$, from which the A-R Conjecture follows. But (6) is not true in general, for instance, if F $= G^4$ and $p = 7$, then F is trivial but F' is elusive (see Ex.5.6(2)). However, it is not known whether or not (6) is true for all nontrivial monotone properties of G^n.

Exercise 5.6

1. Suppose F is a property of G^p where p is a prime. Prove that if $X \in F$ is such that $|\text{Aut } X| = 2p$, then $|\text{Orb}(X)| \equiv (p-1)/2 \mod p$.
2. Show that every nontrivial monotone property F of G^p, where $p < 13$ is a prime, is elusive (Kleitman and Kwiatkowski [80]; Yap [84]).

3.* Does there exist a nontrivial property $F \subset G^n$ such that $F \neq \{O_n\}$ is elusive whereas $F - \{O_n\}$ is non-elusive ?

4. Applying Corollary 6.4, show that if $F \subset G^6$ is a nontrivial monotone property which is not elusive then either $O_1 \cup K_5 \in F$, $O_1 + C_5 \in F$, or $S_6 \in F$.

5.† Prove that every nontrivial monotone property $F \subset G^6$ is elusive.

6. Applying Corollary 6.4, show that if $F \subset G^8$ is a nontrivial, non-elusive monotone property, then either $O_1 \cup K_7 \in F$ or $S_8 \in F$ (but not both).

7. A Counter-example to the Rivest-Vuillemin Conjecture

Rivest and Vuillemin [75] generalized the Aanderra-Rosenberg Conjecture as follows.

R-V Conjecture Let F be a property of T. If the group $\Gamma(F)$ acts transitively on T, $\phi \in F$ and $T \notin F$, then F is elusive.

A counter-example to this conjecture was produced by Illies [78]. We now reproduce this counter-example here.

Let $T = \{1,2,\dots,11,12\}$ and let Γ be cyclic group of order 12 generated by $(1,2,\dots,11,12)$. Let $F = \{\phi, [1], [1,4], [1,5], [1,4,7],$

[1,5,9], [1,4,7,10]} where, say, [1,4] = $\{\{g(1), g(4)\} \mid g \in \Gamma\}$. It is clear that F is invariant under Γ and that Γ (= $\Gamma(F)$) is transitive on T, $\phi \in F$ and $T \notin F$. Thus all the conditions given in the R-V Conjecture are satisfied. However, we shall show that F is not elusive.

We first introduce a shorthand notation as follows : a bracketed leaf $(j_1,\ldots,j_s)$ stands for

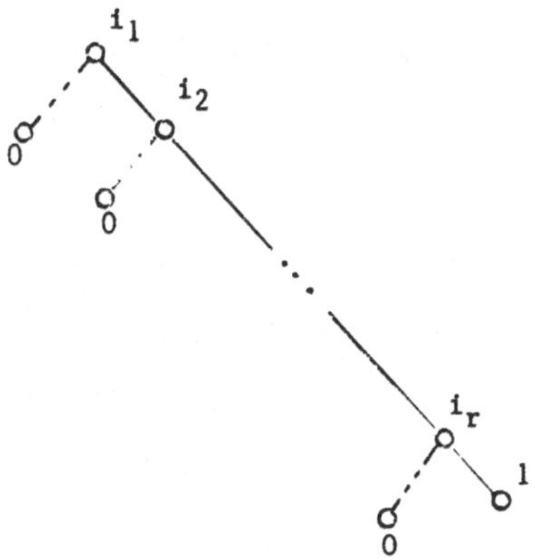

Figure 5.11

where $\{i_1,\ldots,i_r\} = \{i \in \{1,2,\ldots,12\} \mid i \neq j_1,\ldots,j_s$ and i is not a label on the path from the root to the leaf $(j_1,\ldots,j_s)\}$.

To show that the property F given above is not elusive, we may assume, without loss of generality, that the first probe by Algy is 1. Suppose H is the hypothetical set. If $1 \in H$, then 2, 3, 6, 8, 11 and 12 are critical elements and thus should be probed first. It is clear that 2, 3, 6, 8, 11, 12 $\notin$ H otherwise Algy wins the game straightaway.

Suppose Algy's next probe is 4. If $4 \in H$, then $5 \notin H$, and similarly $9 \notin H$. Finally, whether the remaining two elements 7 and 10 belong to H or not is immaterial. On the other hand, if $4 \notin H$, then Algy's next probe is 10. If $10 \in H$, then 5, 9 $\notin$ H, and the remaining element 7 is immaterial; if $10 \notin H$, then $7 \notin H$ and the remaining elements 5 and 9 are immaterial.

The above argument is shown in the right hand branch of the

decision-tree below. The sets {1,4}, {1,4,7}, {1,4,10} and {1,4,7,10} are the sets in F containing in the branch with the leaf (7,10). It can be verified that the depth of this decision-tree is 11. For convenience of reading, the readers may turn it 45° clockwise.

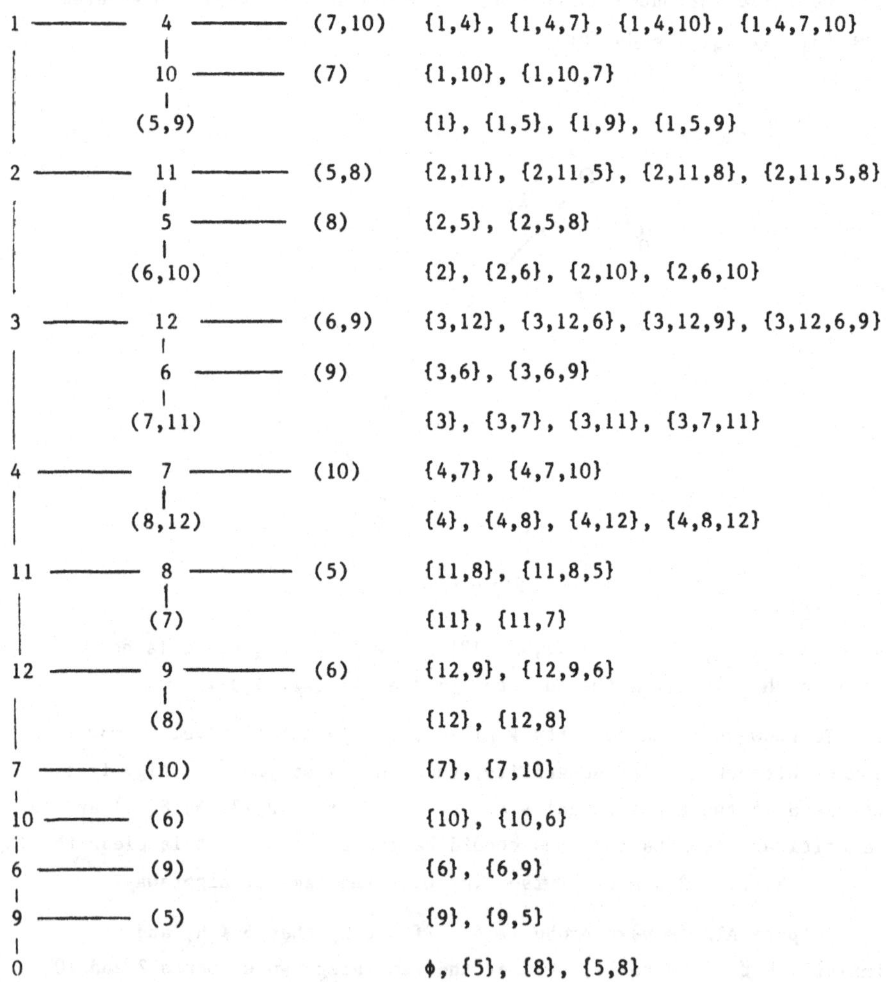

Figure 5.12

Remarks. The set F given in the above counter-example to the R-V Conjecture is not monotone. To revive the R-V conjecture, we may impose an additional condition on F, namely, we require that F be monotone. This new conjecture is called the Generalized Aanderaa-Rosenberg Conjecture.

Generalized Aanderaa-Rosenberg Conjecture Let F be a nontrivial monotone property of T. If the group $\Gamma(F)$ acts transitively on T, and F is monotone, then F is elusive.

Exercise 5.7

1.* Does there exist a nontrivial monotone property which is non-elusive?

8. A lower bound for the computational complexity of graphs properties

To conclude this chapter, we shall apply some theorems on packing of graphs from Chapter 4 to obtain a lower bound for the computational complexity of general graph properties.

Theorem 8.1 (Bollobás and Eldridge [78]) If $P \in \mathcal{G}^n$ is nontrivial, then $c(P) \geq 2n - 4$.

Proof. Since P is nontrivial, by looking at the diagram of P and by the fact that $c(P) = c(\mathcal{G}^n - P)$, we can assume that P has a minimal element $G \neq O_n$. If $e(G) \geq 2n - 4$, then it is clear that $c(P) \geq 2n - 4$. So we may assume that $e(G) < 2n - 4$.

Case 1. $\Delta(G) = n - 1$.

If $G \neq S_n$, then the Constructor picks a vertex v and answers "edge" to any probe vw and "non-edge" to the other $n - 3$ probes ab where $a, b \neq v$. Let H be the pregraph obtained at this stage and let $\alpha = d_H(v)$. Since $e(G - v) < (2n - 4) - (n - 1) = n - 3$ and $e(H - v) = n - 3$, by Corollary 4.3 (ii) of Chapter 4, there is a packing of $G - v$ and $H - v$ and so H can be extended to G in such a way that $d_G(v) = n - 1$. Hence the Constructor can now fix a copy of G with $d_G(v) = n - 1$ and answers "edge" to any further probe cd if and only if cd $\in$

$E(G)$. In this way Algy requires to make a total of at least $(\alpha + n - 3) + (e(G) - \alpha) \geq 2n - 4$ probes. Hence if $c(P) < 2n - 4$, then $G = S_n$.

It is easy to show that $P = \{G \in \mathcal{G}^n \mid \Delta(G) = n - 1\}$ is elusive. Thus we may assume that there is a graph $J \notin P$ with $\Delta(J) = n - 1$. Choose such a graph J so that $e(J)$ is minimum. By the previous argument, if $c(P) < 2n - 4$, then $e(J) < 2n - 4$. Also since $c(P) = c(\mathcal{G}^n - P)$, by the above argument, we have $J = S_n$, a contradiction to the assumption that $J \notin P$ and $S_n \in P$. Hence $c(P) \geq 2n - 4$.

Case 2. $\Delta(G) < n - 1$ and $e(G) \geq n - 2$.

In this case the Constructor answers "non-edge" to the first $m = 2n - 4 - e(G) \leq n - 2$ probes. By Corollary 4.3 (ii) of Chapter 4, G can be packed into $G_m = (V, E_m \cup N_m)$. Let G' be a fixed copy of G in $\bar{G}_m$. To any further probe ab, the Constructor now answers "edge" if and only if $ab \in E(G')$. Hence $c(P) \geq m + e(G) = 2n - 4$.

Case 3. $\Delta(G) < n - 1$ and $e(G) < n - 2$.

If G has an isolated vertex, we can proceed as in Case 2. (We have a packing of G into G_m by mapping an isolated vertex of G to a vertex of valency $n - 1$ in G_m if $\Delta(G_m) = n - 1$.) So we may assume that G has no isolated vertex and hence $e(G) \geq \frac{n}{2}$. Let T be a tree-component of G. Then $|T| \leq n/3$ (because $e(G) \leq n - 3$). Let v be an end-vertex of T and let u be the neighbour of v. Then $d(u) \leq \frac{n}{3} - 1$. If Algy never probes all the places incident with one particular vertex during the first $m = 2n - 4 - e(G)$ probes, we can proceed as in Case 2. However, if Algy does probe all the places incident with one particular vertex, w say, then to the last probe wz, the Constructor gives an "edge" (to all the other $k \leq m - 1$ probes, the Constructor gives a "non-edge"). We note that the valency of u in $G - v$ is at most $\frac{n}{3} - 2 \leq \frac{1}{2}\{(n - 1) - 2\}$ and the valency of z in $G_m - w$ (where $G_m = (V, E_m \cup N_m)$) is at most $m - (n - 1) = n - 3 - e(G) \leq \frac{n}{2} - 3 \leq \frac{1}{2}\{(n - 1) - 2\}$. Moreover, $e(G - v) + e(G_m - w) \leq (e(G) - 1) + (2n - 4 - e(G) - (n - 1)) \leq n - 4 \leq \frac{3}{2}\{(n - 1) - 2\}$. By Ex.4.4 (1), there is a packing σ of $G - v$ into $G_m - w$ such that $\sigma(u) = z$ and thus σ is an embedding of G into $\bar{G}_m + wz$. This forces Algy to make a total of at least $2n - 4$ probes and so $c(P) \geq 2n - 4$. //

Remarks. The above lower bound probably is not best possible. However, the technique applied in the proof indicates that to obtain a better lower bound for the computational complexity of general graph properties will not be easy.

Let $D = (V,A)$ be a digraph such that there is at most one arc joining two distinct vertices of D. The computational complexity of a digraph property is slightly different from that of a graph property. For a digraph property, to any probe ab by Algy, the Constructor has three choices : either $(a,b) \in A$ or $(b,a) \in A$ or none of (a,b) and (b,a) belongs to A.

Bollobás and Eldridge [78] also studied the lower bound for some computational complexity of digraph properties. We include some of their results as exercises here.

Exercise 5.9

1. Prove that $P = \{G \in G^n \mid \Delta(G) = n - 1\}$ is elusive.

2. Suppose P is the property that consists of all digraphs of order n having a sink. Prove that $c(P) = 2n - [\log_2 n] - 2$ (Bollobás and Eldridge [78]; see also Bollobás [78;p.430]).

3. Suppose P is a nontrivial monotone property of digraphs of order $n > 8$. Prove that $c(P) > 2n - [\log_2 n] - 2$ (Bollobás and Eldridge [78]).

REFERENCES

M. R. Best, P. van Emde Boas and H. W. Lenstra Jr., A sharpened version of the Aanderaa-Rosenberg conjecture, Math. Centrum, Amsterdam, 1974.

B. Bollobás, Complete subgraphs are elusive, J. Combin. Theory, Ser.B, 21 (1976), 1-7.

———, Extremal Graph Theory, Academic Press, London, 1978.

B. Bollobás and S. E. Eldridge, Packings of graphs and applications to computational complexity, J. Combin. Theory, Ser.B, 25 (1978), 105-124.

G. J. Fischer, Computer recognition and extraction of planar graphs from the incidence matrix, I.E.E.E. Trans. on Circuit Theory, Vol. CT - 13, No.2 (June, 1966), 154-163.

R. C. Holt and E. M. Reingold, On the time required to detect cycles and connectivity in graphs, Math. Systems Theory 6 (1972), 103-106.

J. Hopcroft and R. Tarjan, Efficient planarity testing, TR73 - 165, Department of Computer Science, Cornell University, April, 1973.

N. Illies, A counter-example to the Generalized Aanderaa-Rosenberg Conjecture, Information Proceeding Letters, Vol.7, No.3, (1978), 154-155.

J. Kahn, M. Saks and D. Sturtevant, A topological approach to evasiveness, Combinatorica 4 (1984), 297-306.

D. Kirkpatrick, Determining graph properties from matrix representations, Proc. 6th SIGACT Conference Seattle (1974), 84-90.

D. Kleitman and D. J. Kwiatkowski, Further results on the Aanderaa-Rosenberg conjecture, J. Combin. Theory, Ser.B, 28 (1980), 85-95.

R. J. Lipton and L. Snyder, On the Aanderaa-Rosenberg conjecture, SIGACT News 6 (Jan., 1974), 30-31.

E. C. Milner and D. J. A. Welsh, On the computational complexity of graph theoretical properties, University of Calgary Research Paper No.232, June, 1974.

———, On the computational complexity of graph theoretical properties, Proc. Fifth British Combinatorial Conf. (Eds., C. St. J. A. Nash-Williams and J. Sheehan), Congre. Numer. XV (1975), 471-487.

R. L. Rivest and J. Vuillemin, On the time required to recognize properties of graphs from their adjacency matrices (Revised), UC Berkeley Electronics Research Laboratory, Memorandum No. ERL-MA76, November, 1974.

———, A generalization and proof of the Aanderaa-Rosenberg conjecture, Proc. of Seventh Annual ACM Symposium on Theory of Computing (1975), 6-11.

———, On recognizing graph properties from adjacency matrices, Theor. Computer Science 3 (1976), 371-384.

A. L. Rosenberg, On the time required to recognize properties of graphs: a problem, SIGACT News 5 (Oct., 1973), 15-16.

R. Tarjan, Depth-first search and linear graph algorithms, SIAM J. Comput., Vol.11, No.2 (June, 1972), 146-159.

H. P. Yap, Computational complexity of graph properties, Graph Theory, Singapore 1983 (Eds., K. M. Koh and H. P. Yap), Springer-Verlag Lecture Notes in Mathematics 1073 (1984), 35-54.

INDEX OF SUBJECTS

INDEX OF NOTATION